Aircraft Icing

OTHER BOOKS IN THE PRACTICAL FLYING SERIES

Aircraft Icing
A Pilot's Guide

Terry T. Lankford

McGraw-Hill

New York San Francisco Washington, D.C. Auckland Bogotá
Caracas Lisbon London Madrid Mexico City Milan
Montreal New Delhi San Juan Singapore
Sydney Tokyo Toronto

Library of Congress Cataloging-in-Publication Data

Lankford, Terry T.
    Aircraft icing / a pilot's guide / Terry T. Lankford.
        p. cm.
    Includes index.
    ISBN 0-07-134139-0
    1. Airplanes—Ice prevention.   2. Airplanes—Cold weather
operation. I. Title.
TL557.I3L35 1999
629.132'5214—dc21

                                         99-41781
                                             CIP

McGraw-Hill

A Division of The **McGraw·Hill** Companies

1 2 3 4 5 6 7 8 9 0 AGM/AGM 9 0 4 3 2 1 0 9

ISBN 0-07-134139-0

The sponsoring editor for this book was Shelley Ingram Carr, the editing supervisor was Caroline Levine, and the production supervisor was Maureen Harper. It was set in Slimbach per the PFS2 design by Michele Pridmore of McGraw-Hill's Professional Group Composition Unit, in Hightstown, NJ.

Printed and bound by Quebecor/Martinsburg.

 This book is printed on recycled, acid-free paper containing a minimum of 50 percent recycled, de-inked fiber.

McGraw-Hill books are available at special quantity discounts to use as premiums and sales promotions, or for use in corporate training programs. For more information, please write to the Director of Special Sales, McGraw-Hill, 11 West 19th Street, New York, NY 10011. Or contact your local bookstore.

Contents

Introduction

Weather affects a pilot's flying activity more than any other physical factor. It is the most difficult and least understood subject in the aviation training curriculum. Weather training for pilots typically consists of bare bones, whereas weather-related accidents—including aircraft icing—remain relatively unchanged.

Over 15 percent of weather-related accidents can be attributed to icing or icing phenomena—carburetor/induction-system icing, freezing precipitation, airframe icing, and snow- or ice-covered surfaces. Whereas most icing accidents fall into the carburetor/induction-system category, only about 10 percent result in fatalities. However, all are preventable. Of the freezing-precipitation/airframe-icing accidents, over one-third are fatal. Approximately one-third of snow/icing accidents occur during the takeoff or landing phase. As with low-level wind shear and thunderstorms, avoidance may be the only solution to the icing hazards.

Report after report during the decade of the 1990s, from both government and industry, has recommended improved aviation weather education for pilots, dispatchers, and controllers alike. On September 27, 1997, Federal Aviation Administration (FAA) Administrator Jane F. Garvey issued the FAA's Aviation Weather Policy statement. It reads in part: "The FAA is committed to improving the quality of aviation weather information and the application of that information by pilots, controllers, and dispatchers. The FAA acknowledges that training is a critical component of this objective, enabling the aviation community to make the best use of weather information to make sound operational decisions and to ensure safety and efficiency."

However, to date, with the noted exception of the Aircraft Owners and Pilots Association's (AOPA) Air Safety Foundation, little has been accomplished.

The FAA and the National Aeronautics and Space Administration (NASA), along with industry groups, are working with the Cooperative Program for Operational Meteorology, Education and Training (COMET) to develop an Aviation Safety Program. NASA has awarded COMET a contract to research, develop, prototype, and implement the program. The goal is to provide pilots, dispatchers, controllers, and forecasters with dedicated training in weather knowledge, interpretation, and decision making in order to foster safety and to most effectively use weather information. However, completion of this program is still years away.

According to the FAA: "Pilots should become familiar with the types of weather associated with and conducive to icing and understand how to detect ice forming on the airplane. Pilots should know the adverse effects of icing on aircraft systems, control, and performance. They should also know how to respond to the situation if accidentally caught in icing conditions. A knowledgeable pilot is better prepared to make timely decisions and promptly recognize the factors that can contribute to aircraft icing accidents."

My goal over the last 10 years has been to share the knowledge gained through over 30 years of flying and flight instruction and over 20 years as an FAA Flight Service Station (FSS) controller.

Pilots gain knowledge through training and experience. All too often I have found that much of my knowledge came through experience. Learning by experience can be defined as where the "test comes before the lesson!" To this end, *The Pilot's Guide to Weather Reports, Forecasts, and Flight Planning* will soon be in its third edition, and in 1997, *Cockpit Weather Decisions* was published. This effort is an attempt to consolidate and compile the theory, effects, and strategies available to avoid icing hazards.

In any work of this type, accident statistics and scenarios are an integral part of the text. With respect to accidents, some may say, Why emphasize the negative? My goal, along with AOPA's Air Safety Foundation, and the FAA, is accident prevention. As already mentioned, people learn through either training or experience. I hope, through a review of accidents, incidents, and scenarios, to help prevent pilots from becoming statistics.

While on the subject of accident scenarios, let's concede that hindsite is always 20-20. It is easy to sit back and analyze and criticize someone else's performance and decisions. When you review an incident in this book, or in any other publication or forum for that matter, do not judge or attempt to assign blame. Incident descriptions are not intended to disparage or malign any individual, group, or organization—the sole purpose is illustration. My goal is prevention through education. Therefore, if the reader perceives any judgment or blame in any incident or scenario, it is strictly unintentional.

Throughout this book, various icing encounters in aircraft not certified for flight in icing will be related. No one should imply that I or the FAA condone such operations. Most occurred because of a lack of training and experience. Like accident incidents and scenarios, their purpose is illustration.

From time to time the FAA is accused of having a "tombstone mentality." In other words, people have to die before anything changes. This notion certainly has some basis in fact, but like many generalizations, it is not always true. A case in point is the 1994 Roselawn, Indiana, accident. As we shall see, this tragic accident spurred a number of changes in aircraft design, regulations, and weather forecast products. Some would say that these changes came 68 human lives too late. On the other hand, aircraft have been operating relatively safely in icing conditions since the 1930s. To quote the U.S. Supreme Court, "Safe does not mean risk-free." We all should acknowledge that the FAA must walk a tightrope between safety and overregulation. This will become quite apparent as you read through this book.

Most of the examples and scenarios related in this book describe general aviation aircraft, generally in the class less than 12,500 lb. In the icing arena, these aircraft are the lowest common denominator and typically the most vulnerable. However, the principles apply to all categories of aviation. This book provides information that is relevant to all grades and types of pilots—from student through airline transport, from recreational to "biz-jet"—to understand the icing environment and how to deal with its hazards. Pilots will be able to recognize atmospheric conditions conducive to icing and develop strategies to deal with this phenomenon.

Various government and nongovernment publications—articles, pamphlets, and videos—have attempted to provide icing education for pilots. Most do not relate weather theory and phenomena to actual flight situations. Articles and pamphlets typically can address this issue only in an oversimplified manner. The same may be said about videos.

Technical meteorologic concepts and terms are translated into language that anyone can understand easily. On the other hand, such subjects as supercooled water and large supercooled drops will not be omitted because of their complexity. Discussions include applying icing theory and strategies to both Visual Flight Rule (VFR) and Instrument Flight Rule (IFR) operation, as well as high-level and low-level operations. This is a practical, bottom line approach to interpreting and applying aircraft icing theory, reports, and forecasts to flight operations.

With the wide variety of airplanes, engines, and equipment in use, it is impossible to include every manufacturer's recommended procedures. With respect to deicing and anti-icing equipment, the pilot should always follow the manufacturer's advice.

Color illustrations are nice, but unnecessarily expensive. They would make the cost of this publication prohibitive, with very little added value. I have used narrative to "color" in the black-and-white illustrations and photographs.

Chapter 1 begins with the physics of icing. This chapter discusses the physical processes of ice formation. To apply icing strategies successfully, these principles must be understood thoroughly. Chapter 1 provides the reader with a solid background for information required in subsequent chapters. Newly developed theories about supercooled cloud and drizzle droplets are presented. Along with these new theories, practical techniques are presented to recognize the potential location of supercooled drizzle droplets.

Chapter 2, gets down to the "nuts and bolts" of structural, induction, and ground icing. Here, the reader develops a sound understanding of the terminology used when describing aircraft icing. This is essential in the proper understanding and application of icing reports, including airport condition Notices to Airmen (NOTAMs), and forecasts. The chapter concludes with a discussion of ice protection equipment.

There are many reports available on icing and icing conditions. Chapter 3 explains the apparent and not so apparent indicators of icing conditions. This chapter identifies and explains weather reports, pilot weather reports (PIREPs), and graphic sources of icing information. Chapter 3 teaches techniques needed to translate, interpret, and apply icing observations. (U.S. domestic PIREPs took on an international look in 1998. Therefore, I have expanded this chapter to emphasize these changes.)

As an FSS controller, I have the opportunity to attend an intensive, week-long satellite interpretation course. For those who would like more information, I recommend *An Introduction to Satellite Image Interpretation*, by E. Conway and the Maryland Space Grant Consortium (Johns Hopkins University Press, 1997). In addition, I will be adding a whole chapter on satellite interpretation to the third edition of *The Pilot's Guide to Weather Reports, Forecasts, and Flight Planning*.

The last section of Chapter 3 discusses the NOTAM system as it relates to airport surface conditions.

Chapter 4 continues with techniques needed to translate, interpret, and apply icing forecasts. Like weather reports, there are numerous icing forecasts. Some are specific icing forecasts, whereas others imply icing potential. All are of some value to the pilot or dispatcher. The chapter concludes with future products and developments in icing forecasts.

Chapter 5 begins with the regulations that apply to aircraft certification in known icing conditions. Now the knowledge of preceding chapters can be applied to flight planning. The chapter proceeds through the collection of icing data and then moves to flight preparation. In the flight-preparation sections I review aircraft and engine considerations as they apply to icing conditions and cold weather operations and then move to pilot and passenger considerations. The chapter concludes with a discussion of risk assessment and management as they relate to flight in icing and cold weather conditions.

With the weather briefing in hand and aircraft, pilot, and passengers prepared for flight, Chapter 6 examines cockpit strategies. The chapter discusses practical procedures to deal with icing conditions throughout the flight—from engine start to taxi through postflight considerations. The final section of Chapter 6 takes the reader through various flight scenarios using an actual icing event. This section looks at both VFR and IFR flights and options for aircraft with and without ice protection equipment.

A glossary is included. The glossary serves as a "quick reference" for terms and concepts associated with aircraft icing.

The chapters herein, hopefully combined with a little humor, explain icing theory and phenomena that apply to aviation. Armed with this knowledge and a commitment not to push the weather, our aircraft, or ourselves, we can realistically achieve a growing safety record in the twenty-first century.

I am greatly indebted to many people, too numerous to be fully listed individually, for their generous help, guidance, and advice. Among them are the meteorologists of the National Weather Service's Aviation Weather Center in Kansas City, Missouri, and members of National Weather Association's (NWA) Aviation Meteorology Committee. I also would like to thank the Aircraft Owners and Pilots Association's Air Safety Foundation, BF Goodrich Aerospace, Ice Protection Systems Division, and the Commander Aircraft Company of Bethany, Oklahoma, for their generous assistance. Finally, there are the pilots and Flight Service Station controllers whom I have been privileged to know and who have allowed me to assist them and in turn provided me with the best aviation weather education possible. This book is dedicated to these people.

The Physics of Icing

The planet Earth, with a thin layer of gas completely covering it, is unique in the solar system. Earth's atmosphere contains about 78 percent nitrogen, 21 percent oxygen, and lesser amounts of argon, carbon dioxide, and other gases—including water vapor. Other unique properties of Earth are temperature and pressure, whose average at the surface allows water to exist as a liquid. The lower layer of the atmosphere, the troposphere, contains about three-quarters of the atmosphere by weight and almost all the water vapor. Nearly all clouds, weather, and icing occur in this layer.

An atmospheric property essential to an examination of the physics of icing is the fact that water can exist as a liquid at temperatures below freezing. In fact, a common misconception, perpetuated by the National Weather Service (NWS) and Federal Aviation Administration (FAA), is that water freezes at 0°C. Water melts at 0°C, which properly should be termed the *melting level.* More technical meteorologic publications and some NWS products do refer to the 0° *isotherm*—line of equal temperature—as the melting level. With this clearly understood, we can move on, using the less accurate but generic term *freezing level.* The significance of this fact is that water can exist in liquid form at temperatures less than 0°C.

FACT

It is a common misperception that water freezes at 0°C. Water *melts* at 0°C.

When this occurs, it is termed *supercooled,* a word that will be used continuously throughout this book.

A pilot's goal when dealing with icing is to avoid, minimize, or escape it. Avoidance is the preferable solution, especially for aircraft not certified for flight into icing conditions. Should a pilot enter, and expect to traverse, an icing layer, the next strategy is to minimize exposure. Once icing is encountered, the final solution—no pun intended—is to escape.

To employ an icing strategy effectively, a pilot must have a thorough knowledge of the atmospheric icing environment. This chapter begins with a discussion of the properties that affect atmospheric icing. These include temperature, moisture, vertical motion, stability, and wind shear. The section on moisture discusses the change of state (i.e., change of water from solid to liquid to vapor). The final sections of the chapter examine the effects of clouds, precipitation and frost, and supercooled large droplets on aircraft icing.

Accretion—the deposit—of ice on aircraft surfaces in flight is a result of the tendency of cloud droplets to remain in a liquid state at temperatures below freezing. The amount and shape of ice depend on temperature, cloud liquid water content, droplet size, aircraft speed, and the horizontal extent of the icing condition.

Atmospheric Properties Affecting Icing

For this discussion, a *property* is a characteristic trait or peculiarity. In specific cases, aircraft icing results from the interaction of specific properties. Of interest to this discussion are the atmospheric properties of temperature, moisture, pressure, and density. (Temperature and moisture are the most significant to the icing story. Therefore, they are discussed in detail in the next two sections.) Certain combinations of these properties result in the formation of clouds, precipitation, and frost. Structural icing in flight occurs only in visible moisture—clouds and precipitation. However, ground icing and carburetor icing often occur in clear air. Aircraft icing can affect both Visual Flight Rule (VFR) and Instrument Flight Rule (IFR) operations.

Aircraft icing is a complex relationship between medium-cold temperatures and medium vertical motions—discussed later in this chapter. According to Don McCann, meteorologist at the NWS's Aviation Weather Center (AWC) in Kansas City, "Who would have thought that such mediocrity could be so dangerous?"

Pressure is defined as force per unit area. The atmosphere exerts pressure. At sea level, atmospheric pressure is approximate 14.7 lb/in.^{2}. (We do not experience this pressure directly because our bodies exert an opposite and equal force. However, we do feel pressure changes in our ears, which must be cleared periodically with changes in pressure or altitude.)

Atmospheric pressure is the weight of all the air molecules above a specific point on Earth. This can be illustrated by an imaginary column of bricks. Assume that there are 10 bricks in the column, each weighing 1 lb. If we weigh the bottom of the stack, the scale reads 10 lb. However, if we weigh the stack at the fifth brick, the scale reads 5 lb; at the first brick, 1 lb. Atmospheric pressure behaves in the same way.

For aviation purposes, pilots commonly relate atmospheric pressure to inches of mercury altimeter setting, or pressure in millibars. Inches of mercury (in. Hg), however, is not a direct expression of force per unit area. The international unit of atmospheric pressure is the hectopascal (hPa), which is equivalent to the millibar (mb).

Like temperature, pressure has a standard. At sea level, it is 29.92 in. Hg, or 1013.2 hPa (1013.2 mb). And guess what? For every level there is a standard temperature and pressure in the standard atmosphere.

In aviation weather, one often sees references to a constant-pressure surface. Layers in the atmosphere are described in this manner. Aviators most often refer to the 850-, 700-, 500-, 300-, and 200-mb constant-pressure surfaces. These standard constant-pressure surfaces occur at approximately 5000, 10,000, 18,000, 30,000, and 39,000 ft, respectively.

The atmosphere, a gas, is compressible. That is, the weight of air molecules above a certain level compresses, or pushes together, the molecules below. Pressure decreases with height, but not at a constant rate. Half the atmosphere by weight exists below 18,000 ft—the 500-mb level.

Although temperature and pressure are important atmospheric properties, to understand the "mechanics" of weather, a knowledge of density is required.

Density is the weight of air per unit volume, often expressed as pounds per cubic foot (lb/ft^3) or grams per cubic meter (g/m^3). Occasionally, one hears the question: "Which weighs less, a pound of dry air or a pound of moist air?" This is the same type question as "Which weighs more, a pound of lead or a pound of feathers?" The answer to both is that they weigh the same—a pound is a pound. However, the *volume* of a pound of dry air compared with a pound of moist air, like the volume of a pound of lead compared with a pound of feathers, is different. These units are not really important to this discussion, but the fact that atmospheric density varies both horizontally and vertically *is* important.

As pilots fly higher in the atmosphere, atmospheric density decreases. As noted earlier, density is weight, and the higher one goes, the fewer molecules there are above; therefore, the air is less dense.

Although pressure is important in the density equation, temperature is often the most significant factor. Typically, the higher the temperature, the lower is the density of the air. Remembering that density is the weight of molecules, at higher temperatures there are fewer molecules per unit volume, pressure remaining constant.

The final factor in air density is moisture. The higher the moisture content of the air, the lower is its density. Why? Water molecules weigh less than air molecules. This is not a major factor, however. What we need to remember is that when the air is humid, air density is less than when conditions are dry.

Temperature

Temperature is a measure of the average speed of molecules. But back up a moment. Heat is the total energy of the motion of molecules; it is a form of energy and thus has the ability to do work. Temperature therefore can be related to the amount of energy available in the atmosphere. And energy drives the weather. Recall the TV weathercaster describing the energy of a particular weather system.

The amount of heat required to raise the temperature of air, or the amount of heat lost when cooled, is known as *heat capacity*. The fact that it is called *heat capacity* is not important. What is important is that energy is released or absorbed during the process. How this relates to aviation weather, and specifically to aircraft icing, will become apparent in subsequent chapters.

> **HEAT CAPACITY**
> The amount of heat required to raise the temperature of air. Or, the amount of heat lost when air is cooled.

Temperature has a direct relationship with the amount of water vapor the air can hold—cool temperature, small amount of water vapor, warm temperature, large amount of water vapor.

The international standard for temperature measurement is degrees Celsius. Winds aloft forecasts have always used this scale. Since July 1, 1996, with the adoption of the METAR (Aviation Routine Weather Report) weather code, the NWS and the FAA have used degrees Celsius exclusively as the temperature standard for aviation in the United States. Pilot weather reports (PIREPs) also use this standard. (More about PIREPs in subsequent chapters.) Although local government and private weather systems doggedly hang onto the old domestic weather report sequence and Fahrenheit scale, all temperatures in this book are in degrees Celsius unless otherwise indicated. (As with the airspace reclassification of 1993, the METAR/TAF codes of 1996, and the change to an international-style PIREP code in 1998, we might as well get used to using degrees Celsius.)

Icing effects and the resulting loss of aircraft performance are directly related to temperature. At cold temperatures, colder than $-15°C$, water freezes quickly on leading edges but does not significantly change the shape of the airfoil. At warm temperatures, greater than about $-2°C$, water freezes slowly, coating surfaces evenly with a glaze, resulting in minimal impact on performance. However, at temperatures between about -4 and $-10°C$, supercooled water does not freeze fast enough to remain on the leading edge; it runs back over the surface and then freezes. Ice accumulation is uneven, causing horns and ridges to form on the wing, disrupting airflow, and significantly degrading performance.

Moisture

If we think of heat as the energy that drives the "weather machine," then moisture, in the form of water vapor, provides the fuel. Without moisture, there would be no clouds or precipitation; in other words, there would be no weather or icing as we know it.

Moisture in the atmosphere occurs in the form of ice crystals (a solid), water (a liquid), or water vapor (a gas). Clouds and precipitation are made up of ice crystals or liquid water. Precipitation is any form of water (solid or liquid) that falls to Earth's surface. Water vapor is invisible, suspended in the air. The effects of water vapor are important phenomena in meteorology and aviation weather.

Because air is a mixture of gases, it has no inherent capacity to hold moisture. The space available in a particular sample of air determines the amount of water vapor molecules it can hold. Since the molecules in warm air are farther apart, warm air can hold more water vapor than cold air—an important fact, as we shall see.

When relative humidity reaches 100 percent, the air is *saturated*. This means that the air can no longer hold any additional water in the form of vapor. Should any more water vapor be added, or should the air be cooled to a lower temperature, condensation occurs in the form of clouds or precipitation—visible moisture that at the right temperature produces structural icing.

Dew point is the temperature to which air must be cooled, water vapor remaining constant, for the air to become saturated. The dew point temperature is usually available in METAR and SPECI (Aviation Selected Special Weather Report) reports, along with air temperature, expressed in degrees Celsius.

Energy, in the form of heat, is required to change water from a solid (ice) to a liquid and from a liquid to a gas (water vapor) (Fig. 1-1). The amount of heat exchanged (absorbed or released) is called *latent heat*. When ice melts, heat is absorbed. The heat required must be supplied from somewhere. It comes from the surrounding environment. In the process of freezing, heat is released to the surrounding environment. When water evaporates (changes

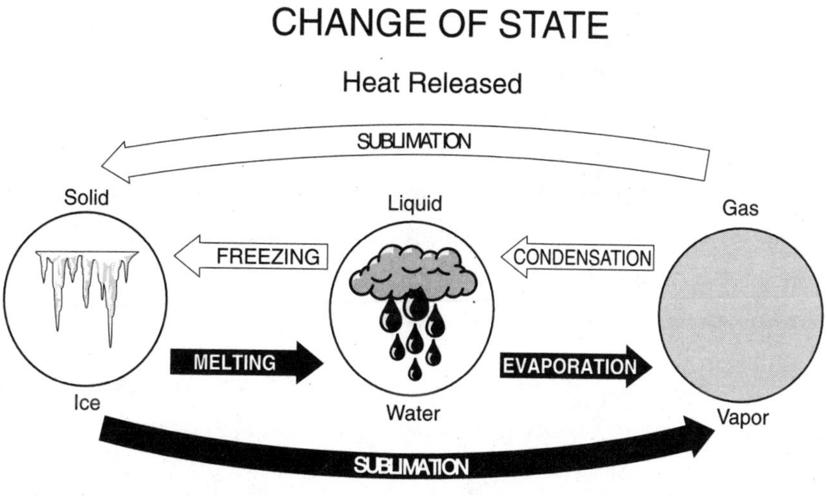

CHANGE OF STATE

Heat Released

Fig. 1-1. During the process of change of state, large amounts of energy, in the form of latent heat, are absorbed from, or released into, the atmosphere.

to water vapor), an even greater amount of heat is absorbed—without any change in temperature. The heat is supplied mostly by the liquid, with a smaller portion coming from the surrounding environment. A primary source of energy in the weather machine comes from the heat released from condensation—the *latent heat of condensation.*

Can water vapor change to ice without going through the liquid stage? Yes. Additionally, ice can evaporate, bypassing the liquid stage. This process is called *sublimation.* For example, an aircraft with structural ice will lose the ice once it is out of the icing environment, even though the temperature remains below freezing. In the process of frost formation, water vapor in the atmosphere sublimates directly to ice. In this process of sublimation, a large amount of heat is absorbed or released.

Melting, freezing, evaporation, condensation, and sublimation are important factors in aircraft icing. When ice melts heat is absorbed. To melt a gram of ice, 80 calories are required. (A *calorie* is the unit of heat required to raise the temperature of one gram of water one degree Celsius.) This process is known as the *latent heat of fusion.* The heat required must be supplied from somewhere. It comes from the surrounding environment. In the process of freezing, heat is released to the surrounding environment.

When water evaporates (changes to water vapor), an even greater amount of heat is absorbed—without any change in temperature. Five hundred and eighty calories per gram are required to condense water vapor to liquid. This is known as the *latent heat of vaporization.* Again, the heat is supplied mostly by the liquid, with a smaller portion coming from the surrounding environment. We can all relate to the latent heat of evaporation. What happens when we get out of a swimming pool? As the water evaporates, our skin cools.

When water vapor condenses (changes to liquid), heat is transferred, or released, to the environment. A primary source of energy in the weather machine comes from the heat released from condensation—the *latent heat of condensation.*

Recall that water vapor can change to ice without going through the liquid stage, bypassing the liquid stage, through sublimation. In the process of frost formation, water vapor in the atmosphere sublimates directly to ice. In this process of sublimation—*latent heat of sublimation*—a large amount of heat is absorbed or released. It approximately equals the sum of the latent heats of fusion and vaporization.

Melting of ice, evaporation of water, and sublimation of ice to vapor are cooling processes. Water during this change of state takes heat from the surrounding environment. Freezing, condensation, and sublimation of water vapor to ice are warming processes. Heat is added to the environment. We will refer to this process of adding heat to the environment in subsequent chapters.

When temperatures during the entire process of cloud formation are below freezing, vapor generally sublimates directly as snow or ice crystals. When temperature during the process changes from above to below freezing, clouds and precipitation may be in a variety of forms, including ice crystals, snow, ice pellets, hail, snow grains, and supercooled water.

Atmospheric Phenomena Related to Icing

Specific atmospheric phenomena connected with aircraft icing are

- Vertical motion
- Stability
- Wind shear

In addition to the atmospheric properties of temperature, moisture, pressure, and density, specific atmospheric phenomena are directly or indirectly connected with aircraft icing. These phenomena are vertical motion, stability, and wind shear.

Vertical motion in the atmosphere has a direct effect on the icing environment. It causes clouds and precipitation to develop or dissipate. The strength of vertical motion affects the characteristics of cloud and precipitation particles. These characteristics are directly related to the intensity of icing present in a particular cloud.

Stability has a direct effect on weather in general and a direct effect on icing. We all know that stability is one of the key players in the development of thunderstorms. As we shall see, the type, extent, and severity of icing have a good deal to do with the stability of the atmosphere.

Finally, we have wind shear. Wind shear is most often associated with its hazards to departing and arriving aircraft during convective occurrences (low-level wind shear) and turbulence at high altitude (clear-air turbulence). However, like vertical motion and stability, there is a relationship between severe icing and certain cases of wind shear.

Vertical motion can be produced, enhanced, or dampened by one or all of the following:
- Convergence and divergence
- Frontal lift
- Dry line
- Vorticity
- Upslope
- Pressure systems
- Convection
- Warm- and cold-air advection
- Thunderstorms

Vertical Motion

An understanding of vertical motion (both upward and downward) explains many weather phenomena and aviation weather hazards, including aircraft icing. This section briefly explores the methods employed by nature to move air, and its properties, vertically.

During this discussion it may be useful to refer to Table 1-1. This table indicates whether motion is upward or downward—stability. The table also includes weather products where these phenomena are observed and forecast.

Convergence refers to an inflow or squeezing of the air. On the horizontal, when airflow into an area is greater than the outflow, the air literally piles up. Since the ground prevents the air from going downward, there is only one way left for it to go—up. The bottom line is that an area of convergence is an area of rising air. Convergence can occur aloft over dense, cold air and is not necessarily confined to a layer bounded by the surface. When moisture is adequate and convergence is great enough, condensation occurs.

CONVERGENCE
Inflow or "squeezing" of the air.

Convergence occurs along surface low-pressure troughs and at the centers of low-pressure areas. An area of strong winds blowing into an area of lighter winds causes wind-speed convergence. When this occurs near the surface, the result is a region of rising air.

Table 1-1
Producers of Vertical Motion

PHENOMENON	STABILITY	OBSERVED	FORECAST
Convergence	Up	Surface	850-mb/700-mb SIG WX prog
Divergence	Down	Surface	850-mb/700-mb SIG WX prog
Fronts	Up	Surface	SIG WX prog
Dry line	Up	Surface	
Vorticity	Up/down	500-mb heights/vorticity	500-mb prog heights/vorticity
Orographic	Up/down	Surface	SIG WX prog
High pressure (SFC)	Down	Surface	SIG WX prog
Low pressure (SFC)	Up	Surface	SIG WX prog
Ridge (SFC)	Down	Surface	SIG WX prog
Trough (SFC)	Up	Surface	SIG WX prog
High pressure (aloft)	Down	Constant-pressure charts	Constant-pressure progs
Low pressure (aloft)	Up	Constant-pressure charts	Constant-pressure progs
Ridge (aloft)	Down	Constant-pressure charts	Constant-pressure progs
Trough (aloft)	Up	Constant-pressure charts	Constant-pressure progs
Convection	Up		
Warm-air advection (SFC)	Up	850-mb/700-mb charts	850-mb/700-mb progs
Cold-air advection (SFC)	Down	850-mb/700-mb charts	850-mb/700-mb progs
Warm-air advection (aloft)	Up	300-mb/200-mb charts	300-mb/200-mb progs
Cold-air advection (aloft)	Down	300-mb/200-mb charts	300-mb/200-mb progs
Thunderstorms	Up	Radar summary	SIG WX prog/SVR WX outlook

Divergence is the opposite of convergence. Downward motion (subsidence) of air causes it to spread out at Earth's surface. From a knowledge of lapse rate, it is easy to see that divergence is a drying and stabilizing process. Like convergence, divergence may occur in a layer aloft not extending to the ground. Divergence occurs along high-pressure ridges and at the centers of high-pressure areas. When air near the surface blows from an area of light winds into an area of stronger winds, wind-speed divergence results.

Observed and forecast locations of convergence and divergence can be obtained from the surface analysis chart, from 850- and 700-mb constant-pressure charts, and from the low-altitude significant weather prognostic (prog) chart (SIG WX Prog).

FRONTAL ZONE
A zone of rapid change separating two air masses.

Air masses of different properties (temperature and moisture) do not tend to mix. Differences in temperature, humidity, and wind may change rapidly over short distances. Where there are temperature and moisture differences, there is a difference in density. This zone of rapid change separating two air masses is a *frontal zone,* more commonly referred to as a *front.* In these zones, the less dense air is lifted. This causes vertical motion in the atmosphere. The type of weather produced depends on the stability of the atmosphere.

A *dry line,* or *temperature–dew point front,* marks the boundary between moist, warm air from the Gulf of Mexico and dry, hot air from the southwestern United States. Dry lines usually develop in New Mexico, Texas, and Oklahoma during the summer months. Since the moist air from the Gulf of Mexico is less dense than the dry, hot desert air, it is forced aloft. If the air mass is unstable, thunderstorms and tornadoes develop along the boundary. As shown in Table 1-1, dry lines appear on the surface analysis chart but are not forecast on prog charts.

Anything that spins has vorticity, which includes Earth. *Vorticity* is a mathematical term that refers to the tendency of the air to spin; the faster that air spins, the greater is its vorticity. [Note from Table 1-1 that vorticity can produce either upward (unstable) or downward (stable) vertical motion.] A parcel of air that spins counterclockwise (cyclonically) has positive vorticity; a parcel of air that spins clockwise (anticyclonically) has negative vorticity.

Air moving through a ridge, spinning clockwise, gains anticyclonic relative vorticity. Air moving through a trough, spinning counterclockwise, gains cyclonic relative vorticity. Therefore, there tends to be downward vertical motion in ridge-to-trough flow and upward vertical motion in trough-to-ridge flow. More about this later.

The observed location and forecast position of positive and negative vorticity can be found on 500-mb heights/vorticity charts.

Orographic is a term used to describe the effects caused by terrain, especially mountains. An *orographic effect* is an upslope or a downslope. Besides the adiabatic process, air also can take on the characteristics of the terrain through the process of conduction. That is, air can absorb heat or moisture by direct contact with the surface. Air moving up a slope rises and tends to cool; air moving down a slope sinks and tends to warm.

Like vorticity, orographic effect can be either upward or downward. Its effects can be inferred from the surface analysis chart for current conditions and the significant weather prog for forecast events, with a knowledge of terrain.

Circulation around high-pressure areas produces downward vertical motion and around low pressure produces upward vertical motion. Typically, but not always, high pressure means good weather and low pressure means poor weather.

Downward vertical motion occurs along *high-pressure ridges* (usually referred to as a *ridge*). Conversely, upward vertical motion occurs along *low-pressure troughs* (usually referred to as a *trough*).

Vertical motion also occurs at higher levels. For example, air moving from an upper ridge to a trough produces downward vertical motion; air moving from an upper-level trough to a ridge produces upward vertical motion. The location and forecast position of upper-level ridges and troughs can be found on constant-pressure charts and progs.

Atmospheric convection is the transport of a property vertically. For our purposes, near the surface, convection is caused by surface heating. Surface heating, and the resulting convection, are primary vertical motion producers. Areas of convection cannot be found directly on standard aviation weather charts and progs. However, areas of convection can be inferred in areas of high temperature, especially during afternoon heating.

> **ADVECTION**
> is a term used to describe the movement of an atmospheric property from one region to another. Temperature, moisture, and stability are properties that can be advected.

From the surface to about 10,000 ft, warm-air advection produces upward vertical motion and cold-air advection produces downward vertical motion. As warmer, less dense air moves into an area, it will tend to rise. Warm-air advection causes surface pressures to fall. This results in convergence and upward vertical motion. When cooler, more dense air moves into an area, it tends to sink. Cold-air advection causes surface pressures to rise. This results in divergence and downward vertical motion. Therefore, warm-air advection destabilizes conditions, whereas cold-air advection tends to stabilize the weather at or near the surface. Areas of observed and forecast warm- and cold-air advection can be found on 850- and 700-mb constant-pressure charts and progs.

Above the 500-mb level, the opposite occurs. Cold-air advection destabilizes conditions and warm-air advection stabilizes the atmosphere. Cold-air advection above the 500-mb level decreases the lapse rate. This enhances any convective activity that might develop. Conversely, warm-air advection aloft stabilizes the atmosphere by increasing the lapse rate, thus retarding any convection. Areas of observed and forecast warm- and cold-air advection can be found on 300- and 200-mb constant-pressure charts and progs.

A tremendous vertical motion producer, a *thunderstorm* is a local storm produced by cumulonimbus clouds. The storm itself may be a single cumulonimbus cloud or cell, a cluster of cells, or a line of cumulonimbus clouds that in some cases may extend for several hundred miles. Thunderstorms are always associated with one of the previously mentioned vertical motion producers. Observed thunderstorm activity is found on real-time radar displays and the radar summary chart; forecast locations are depicted on the significant weather prog and severe weather outlook chart and inferred from moisture/stability charts.

Large upward vertical motion in the atmosphere enhances large liquid water content (LWC). However, large upward motion produces drops too large for hazardous ice formation. Slow upward motion does not produce much LWC. Thus droplets are too small for serious ice formations. Medium upward vertical motion, on the order of 10 cm/s, creates both a large cloud LWC and relatively large drops. This environment has the potential for the most severe icing conditions.

Stability

Stability is the tendency of an air mass to remain in equilibrium—its ability to resist displacement from its initial position. If we move a parcel of air and then remove the lifting mechanism, one of several things occurs: The parcel will tend to return to its original position; the parcel will continue to rise without any additional lifting force; the parcel initially will resist upward displacement to a certain point, where it then continues upward spontaneously; or the parcel will remain at the level where the external force ceased. (A parcel is a small volume of air; it retains its composition and does not mix with the surrounding air.) These processes are known as

- Absolute stability

- Absolute instability

- Conditional instability

- Neutral stability

A parcel is *absolutely stable* when it resists vertical displacement, whether saturated or unsaturated. The parcel is always cooler (denser) than the surrounding air, so it wants to sink. Thus vertical motion is impossible unless caused by an external force.

Air is *absolutely unstable* when vertical displacement of a parcel within the layer is spontaneous, whether saturated or unsaturated. If a parcel is lifted, its temperature is always warmer than the surrounding air. The cooler, more dense air surrounding the parcel forces it upward. Vertical motion is spontaneous, and the layer is absolutely unstable.

Let's look at the third case, conditional unstable. *Conditional instability* refers to the structure of a column of air that will produce free convection of a parcel as a result of it becoming saturated when forced upward. *Free convection* means that once saturation occurs, upward movement will continue spontaneously. This is the *level of free convection* (LFC). A lifted parcel is stable to the point where saturation occurs. Below the lifted condensation level, the parcel exhibits absolute stability. However, on saturation, upward displacement becomes spontaneous. The parcel becomes unstable on the condition that it reaches saturation. High moisture content in low levels and dry air aloft favor instability. Conversely, dry air in low levels and high moisture content aloft favor stability.

The concepts of absolute stability and absolute instability are relatively straightforward. Conditional stability is more complex and includes many subclassifications. It depends not only on temperature but also on water vapor distribution.

When a parcel is displaced and remains at rest—even when the displacing force ceases—the layer is *neutrally stable*.

Wind Shear

Wind shear is any change in wind speed or direction, either vertically or horizontally, over a relatively short distance. When wind shear occurs within 2000 ft of the surface, it is referred to as *low-level wind shear* (LLWS). Wind shear is also a significant factor in turbulence and plays a significant part in the formation and intensity of thunderstorms. LLWS, turbulence, and thunderstorms as a separate subject are not topics of this book.

One of the properties of wind shear aloft is to increase vertical motion. Strong, high-level wind shear can significantly enhance thunderstorm and tornado activity. Moderate midlevel wind shear enhances icing potential by allowing larger supercooled droplets to remain suspended.

Although at present pilots have no direct access to midlevel wind shear information, meteorologists take this phenomenon into account when preparing icing forecasts. Any mention of a wind shear environment in the weather synopsis or forecasts should alert the pilot to increased icing potential and severity.

Clouds

Clouds form through a cooling process. The process initiates and then must sustain condensation or sublimation. Cooling processes are adiabatic or diabatic.

The *adiabatic* process cools the air by lifting the parcel through the processes of convection, convergence, or orographic lifting. The *diabatic* process produces a loss of heat. The loss may occur through terrestrial radiation, resulting in fog or low clouds. Conduction through contact with a cold surface may result in dew, frost, or fog. The process may be associated with movement of air across a cold surface—advection. Finally, the process may occur through mixing with colder air. If the mixture has a temperature below its dew point, clouds or fog may form.

There are a number of methods of cloud classification. Clouds may be classified according to appearance, how they are formed, or the

Table 1-2
Cloud Characteristics

NAME	CATEGORY	BASE, FT	HEIGHT	STABILITY	PRECIPITATION	EXTENT/LWC
Cirrus (curly)	Cirroform	ABV 20,000	High	Stable	None	Ice crystals
Cirrostratus	Cirroform	ABV 20,000	High	Stable	None	Ice crystals
Cirrocumulus	Cirroform	ABV 20,000	High	Unstable	None	Ice crystals
Altostratus (Alto-high)	Stratiform	6500 to 20,000	Middle	Stable	RA SN	Widespread—solid and liquid water
Altocumulus	Cumuliform	6500 to 20,000	Middle	Unstable	VIRGA	Limited—solid and liquid water
Stratus (spread out)	Stratiform	SFC-6500	Low	Stable	DZ	Widespread—high LWC
Stratocumulus	Stratiform	SFC-6500	Low	Slightly unstable	DZ –RA	Limited—high LWC
Cumulus (heaped up)	Cumuliform	Near SFC	Vertical development	Unstable	SHRA GS	Limited—high LWC
Nimbostratus (Nimbus-rain)	Stratiform	SFC-6500	Low	Stable	–RA RA	Widespread—high LWC
Cumulonimbus	Cumuliform	Near SFC	Vertical development	Unstable	+TSRA +RA GR	Limited—high LWC

height of their bases. The various methods are compared in Table 1-2 and discussed below.

It was not until 1803 that cloud forms were first classified. Luke Howard, an Englishman, divided clouds into three main categories using Latin names:

- *Cirrus*—meaning "curly"
- *Stratus*—meaning "spread out"
- *Cumulus*—meaning "heaped up"

Two prefixes/suffixes may be added:

- *Alto*—meaning "high"
- *Nimbo*—meaning "rain"

Today, meteorologists divide clouds into four main groups based on cloud height.

- *Low clouds* (bases near the surface to about 6500 ft)
- *Middle clouds* (bases from 6500 to 20,000 ft)
- *High clouds* (based at or above 20,000 ft)
- *Clouds with vertical development* (based near the surface, tops of cirrus)

Sometimes clouds are classified into one of two general classifications: stratiform and cumuliform. *Stratiform* describes clouds of extensive horizontal development, associated with a stable air mass. Stratiform clouds consist of small water droplets. The following cloud types are classified as stratiform:

- Stratus
- Stratocumulus
- Nimbostratus
- Altostratus
- Cirrostratus

Cumuliform describes clouds that are characterized by vertical development in the form of rising mounds, domes, or towers, associated with an unstable air mass. Because of upward moving currents, cumuliform clouds can support large water droplets. In the case of cumulonimbus clouds, updrafts can support hail. The following cloud types are classified as cumuliform:

- Altocumulus

- Cirrocumulus

- Cumulus

- Cumulonimbus

A third generic cloud type, not included in either Howard's Latin classification or the modern method based on height, may be used. It describes the entire group of high clouds—cirroform. *Cirroform* is often used during pilot weather briefings to translate one or all of the high-cloud types: cirrus, cirrostratus, or cirrocumulus.

Stratus clouds indicate a stable air mass. Precipitation, when it occurs, is usually light, often in the form of drizzle. A stratus cloud on the ground is called *fog*. Freezing fog consists of supercooled droplets that freeze on contact with exposed objects. This may result in the formation of icing.

Supercooled fog is identical to a supercooled cloud except for altitude. Supercooled fog can have the same effect as freezing drizzle. Prior to 1996, *ice fog* was used interchangeably with *supercooled fog,* but technically it represented a fog composed of ice crystals. Freezing fog (FZFG) used in METAR/TAF today is any fog consisting predominately of water droplets at temperatures below 0°C, whether it deposits *rime*— a white or milky and opaque granular deposit of ice—or not.

Stratocumulus clouds represent a moist layer with some convection. Stratocumulus clouds may form from the spreading out of cumulus clouds, which indicates decreasing convection. Stratocumulus clouds can develop from stratus clouds with winds of moderate to strong intensity.

Nimbostratus clouds are low clouds, usually uniform and dark gray in color. Nimbostratus clouds usually evolve from altostratus clouds that have thickened and lowered, sometimes with a ragged appearance. This is the ordinary rain cloud that produces light to moderate, steady precipitation.

These lower-level stratiform clouds are important to the icing environment because of their LWC and horizontal extent. Within low-level stratiform clouds, air temperature decreases with altitude, but LWC increases, reaching a maximum at or near the top of the cloud. Droplet diameter also increases with increasing altitude. This helps explain why icing intensity tends to be greatest near cloud tops.

Icing in stratiform clouds is normally in the middle- to lower-level clouds below 10,000 ft. Stratiform clouds are quite stable, leading to the formation of clouds with extensive coverage and multiple layers, with a total thickness on the order of 6500 ft.

Flight in multiple-layer clouds can result in icing conditions of long duration and, form the criteria most often used for design of ice protection systems, for such components as wings, empannage, propeller, and windscreens. The intensity of icing generally ranges from light to moderate, with maximum values occurring in the upper portion of the cloud. The main hazard results from the great horizontal extent of some cloud decks. Stratiform clouds are characterized by moderate LWC and most frequently produce rime icing.

The icing envelope for stratiform icing encounters is shown in Fig. 1-2. In low-level stratiform clouds, encounters with icing rarely occur at temperatures below −18°C. At higher levels, up to 22,000 ft, the lowest icing temperature is −30°C.

Middle clouds fall into two general types: altostratus and altocumulus. Altostratus clouds indicate a stable atmosphere at midlevels. Some altostratus clouds are thin and semitransparent, whereas others are thick

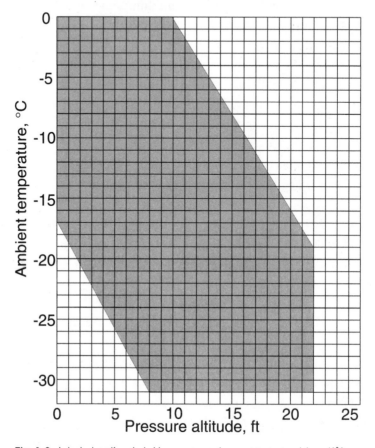

ICING ENCOUNTERS IN STRATIFORM CLOUDS

Fig. 1-2. In low-level stratiform clouds, icing encounters rarely occur at temperatures below −18°C.

enough to hide the sun or moon. Altostratus clouds often indicate the approach of a warm front. These clouds can produce precipitation in the form of rain or snow even heavy snow at times.

Altocumulus clouds indicate vertical motion and instability at midlevels. Altocumulus clouds may be thin, mostly semitransparent. Some altocumulus clouds are thick, developed, and may be associated with other cloud forms. This cloud often signals the approach of a cold front.

Altocumulus castellanus (ACC), a midlevel cloud, indicates moisture and instability at this level. ACC clouds may indicate thunderstorm development. When showers falling from these clouds evaporate before reaching the surface, it is known as *virga.*

Middle-level clouds are important to the icing environment because they typically contain both solid and liquid water content. These clouds may be very thin, but they can extend over areas up to thousands of square miles.

Cumuliform clouds are important to this discussion because of their rapid development and large LWC. As a result of adiabatic lifting, these clouds often contain supercooled droplets. Cumuliform clouds cover less area horizontally than stratiform clouds. Icing is variable and depends on the stage of development of the individual cloud. Although icing may occur at all levels above the freezing level in building cumulus clouds, it is most intense in the updrafts that support larger supercooled water droplets. These clouds are characterized by short-duration exposures to typically high LWC. Maximum water content of a single cell is likely to be at the cloud center. For a single cell, horizontal extent averages from 2 to 6 nautical miles (nmi). However, this will be greater in convective cloud lines or clusters and areas where convective clouds are embedded in stratiform clouds. Icing in these clouds is usually clear—a relatively transparent, homogeneous layer or mass of ice—or a mixture of clear and rime ice.

High clouds are known as cirrus, cirrostratus, and cirrocumulus clouds. Cirrus clouds are composed entirely of ice crystals. Usually the air is so cold that they do not present an icing hazard.

Cirrus clouds often consist of filaments, commonly known as *mares' tails.* Other cirrus clouds are associated with cumulonimbus clouds. A thickening cirrus layer may indicate the approach of a front.

Cirrostratus clouds are sheets or layers of cirrus clouds. Sun or moon halos may occur. When cirrostratus clouds appear within a few hours after cirrus clouds in the middle latitudes, there is a good probability of an approaching front. Cirrocumulus clouds indicate vertical motion at high levels. Cirrus clouds, of themselves, have no significance to low-level flights.

Clouds with vertical development are cumulus and cumulonimbus clouds. Some cumulus clouds describe fair weather. Other cumulus clouds contain considerable vertical development, generally towering. This type precedes the development of cumulonimbus clouds and thunderstorms. METAR reports may contain TCU (towering cumulus) to describe such clouds. This refers to growing cumulus clouds that resemble a cauliflower but with tops that have not yet reached the cirrus level. With cumulus clouds, expect showery weather and often heavy precipitation.

Cumulonimbus clouds exhibit great vertical development, with tops composed, at least in part, of ice crystals. The tops no longer contain the well-defined cauliflower shape of towering cumulus clouds. Cumulonimbus clouds may develop a clearly fibrous (cirroform) top, often anvil-shaped. Regardless of vertical development, a cloud is classified as cumulonimbus only when all or part of its top is transformed, or in the process of transformation, into a cirrus mass. Any cumulonimbus cloud should be considered a thunderstorm with all its ominous implications.

The contractions *TCU* and *CB* appear in the body of METAR reports, and *CB* is used in TAFs to indicate the presence of these clouds.

The icing envelope for cumuliform clouds is somewhat more narrow than for stratiform clouds (Fig. 1-3). The minimum altitude is 4000 ft, and maximum altitude is 24,000 ft. Icing is most frequently encountered (about 40 percent of the time) at ambient temperatures of 0 to $-10°C$. At lower temperatures, the frequency of icing in clouds is much lower, about 6 percent at $-30°C$.

The largest LWCs occur in cumulus and cumulonimbus clouds (generally from 100 to 300 nmi behind a cold front), in maritime air

ICING ENCOUNTERS IN CUMULIFORM CLOUDS

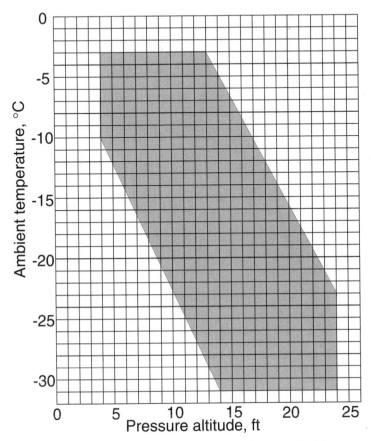

Fig. I-3. With cumuliform clouds, icing encounters typically occur between 4000 and 24,000 ft.

masses, and in areas of orographic lifting. In the United States, supercooled convective clouds rarely occur at altitudes below 3000 ft above ground level (AGL), except in mountainous areas.

The following is background information, containing additional detail for those who would like a better understanding from a more scientific perspective. Cloud LWC is expressed in grams of liquid water per cubic meter of air (g/m³). It includes only the water in supercooled droplet form, not the water in vapor form. Typical values range from 0.1 to 0.8 g/m³ for stratiform clouds and 0.2 to 2.5 g/m³ for cumuliform clouds. Droplet size is typically expressed in micrometers or microns (μm). Any droplet larger than about 100 μm tends to fall as precipitation. Median volumetric diameter (MVD)—average droplet diameter—generally is less than 35 μm, but droplets as large as 100 μm have been reported. Freezing rain may involve droplets as large as 1000 μm (1 mm).

Icing conditions in the United States in winter months become most serious at altitudes of 7000 to 9000 ft AGL. This is where maximum LWC occurs. Above 10,000 ft, maximum LWC appears to decrease in wintertime clouds. A secondary area of concern occurs at altitudes from 4000 to 6000 ft AGL. This layer corresponds to the typical upper limits of the turbulent mixing layer in many wintertime situations. Data show that the most serious icing occurs between 3000 and 5000 ft AGL, with another peak at 8000 ft AGL. Icing occurs less frequently at high altitude—up to 22,000 ft—where minimum observed icing temperature is −30°C. This is due to the fact that LWC decreases with decreasing temperature. When encountered, icing is typically "light" in intensity. The minimum temperature usually referenced for the existence of supercooled water droplets is −40°C. At this temperature, virtually all water is converted to ice crystals. Therefore, icing is not likely to be a factor.

The winter, or cool season, months with large, cold air masses contain much less water vapor due to the lower temperatures. The

THE SEASONS

Over the latitudes of the United States there are, more or less, four seasons. However, a good portion of the country does not have four distinct seasons. It is often useful to relate seasonal weather changes to the winter, cool season and summer, warm season. Winter, cool season weather patterns occur between the months of November and April. Summer, warm season weather patterns occur between the months of May and October. Winter patterns change to summer during May and June; summer patterns change to winter during November and December. Occasionally, weather from one seasonal pattern overlaps into the other. This is especially true during the transition months.

cool air masses create more stable temperature environments leading to the creation of stratiform cloud systems instead of the cumuliform systems of the warmer season. The lower freezing temperatures of winter, however, lead to the increased likelihood of aircraft icing encounters. In addition, the time in cloud formations can be extensive, thus creating a critical icing environment.

EXCEPTIONS

With all this said, there are exceptions. (It has been stated that the first word in the Federal Aviation Regulations is *except*. Well, this applies equally to weather.) For example, take the following pilot weather report:

FAT UUA /OVC CZQ 300040-CZQ 015045/TM 1640/FL220-260/TP NMRS
/IC MOD-SEV MIXED

During the early spring of 1997, California was struck by a series of storms, colloquially known as the "pineapple express." These storms began in the tropical latitudes of Hawaii and brought moist, unstable air to California. In the San Joaquin Valley, numerous air carriers were reporting moderate to severe mixed icing between FL220 and FL260. Typically, this intensity does not occur in the mid-twenties. On this occasion, there was a warm, moist, unstable southwesterly flow. Temperatures for these altitudes were forecast between -18 and $-26°C$. Weather radar was reporting widespread light to moderate rain and scattered areas of heavy rain. Add the orographic uplift of the Sierra Nevada Mountains, which support larger supercooled droplets, and conditions were perfect for this phenomenon to occur. The point is that pilots and forecasters must look carefully at each individual situation to determine the probability and intensity of icing.

The summer, or warm season, months create large, warm air masses that can contain large amounts of water vapor. The high temperatures also can create temperature instability, leading to strong convective updrafts that, in turn, cause cumuliform cloud formations that have relatively high LWC values. These formations can create very severe local icing encounters.

Another factor of the warm season is the height of the freezing level. With the freezing level sometimes as high as 16,000 ft or higher in the southern United States, icing in clouds can extend well into the flight levels. This is especially a factor for pilots of nondeiced turbo-charged airplanes. During winter, these airplanes can quickly climb above the icing threat. However, during the summer, these same altitudes may put the pilot in the middle of an icing layer.

Variations in geography also play a role. For the purposes of this discussion, the country can be divided into three geographic regions: Eastern United States, Plateau, and Pacific Coast. The boundary between the Eastern and Plateau regions is the eastern border of Montana, Wyoming, Colorado, and New Mexico. The boundary between the Plateau and Pacific Coast regions is the eastern border of Washington, Oregon, and California.

> Areas of greatest icing concern in the United States are the Great Lakes, coastal areas, and mountainous regions.

Areas of greatest icing concern in the United States are the Great Lakes, coastal areas, and mountainous regions, although fronts with freezing rain or other icing conditions occur in most areas. The Great Lakes and coastal areas, particularly near the Gulf Stream, enhance the possibility of encountering high values of LWC due to the proximity of large amounts of water vapor and lake effect. Large bodies of water add moisture and heat to the lower layers of a cold air mass. This induces instability, resulting in convection. Forced lifting of air on the windward sides of mountain ranges—orographic effect—causes the air to cool and clouds to form with adequate moisture. If the air is stable, stratiform clouds form. If the air is conditionally unstable, convection will occur and cumuliform clouds will form. The Appalachian and Pacific Coast mountains cause significant orographic effects. A warm, moist maritime influence may greatly enhance this effect along the higher Pacific Coast ranges.

Cloud formation is often cyclic, especially for cumuliform clouds. That is, clouds are in the process of forming and dissipating—a cycle. At the beginning of the cycle there is little vertical motion, thus there is little or no precipitation and only trace to light icing above the freezing level. As the cycle progresses, vertical motion increases, bringing liquid water droplets above the freezing level. As light precipitation begins to fall, light to moderate icing occurs in the lower levels above the freezing level, with greater intensities higher in the cloud. As downdrafts begin, heavy precipitation develops; however, in the cloud, only trace to light icing occurs. Then the process begins all over again. The time span on this process is on the order of an hour.

Precipitation and Frost

Precipitation is any or all of the forms of water particles, whether liquid or solid, that fall from the atmosphere and reach the ground. Precipitation does not include clouds, fog, dew, or frost. Fog, dew, and frost occur as a result of condensation through contact with a cold surface.

FORMS OF PRECIPITATION
- Drizzle
- Rain
- Freezing drizzle and freezing rain
- Snow grains
- Snow
- Snow pellets
- Ice crystals
- Ice pellets
- Hail

Drizzle is very small, numerous, and uniformly dispersed water drops. *Rain* is precipitation in the form of liquid water drops. Neither produces structural icing.

Freezing rain and *freezing drizzle* are caused by liquid precipitation falling from warmer air into air that is at or below freezing. Droplets freeze on impact with a surface below freezing. Structural icing can be expected while flying through freezing precipitation, probably the most dangerous of all icing conditions. It can build hazardous amounts of ice in a few minutes and is extremely difficult to remove. Freezing rain and freezing drizzle can flow back along the aircraft, sometimes beyond the ice-protected areas and can cover the static ports.

Snow grains are small, white, opaque grains of ice, the solid equivalent of drizzle. Since snow grains are already frozen, they typically do not present an icing hazard.

Snow is composed of white or translucent ice crystals, chiefly in complex branched hexagonal form and often integrated into snowflakes. Dry snow does not lead to the formation of aircraft structural ice because the particles are dry and do not adhere to aircraft surfaces, except for heated engine inlets. However, wet snow—snow that contains a great deal of liquid water—produces structural icing.

HOW WARM CAN IT SNOW?

Snow can fall about 1000 ft below the freezing level before melting. Snow often can begin with temperatures of 2°C; it is even possible to see snowflakes at temperatures around 10°C. This occurs only when the air is very dry. As snow falls into above-freezing air, it begins to melt. The water evaporates and cools the air. Evaporation cools the snow, which retards melting. Water vapor is added to the air, which increases the dew point. Finally, the air cools and becomes saturated at 0°C.

Snow pellets—small, white, opaque grains of ice—form when ice crystals fall through supercooled droplets and the surface temperature is at or slightly below freezing. Falling from cumuliform clouds, snow pellets are more prone to cause structural icing than snow grains. Snow pellets are also known a *soft hail* or *graupel*.

Ice crystals may appear suspended and fall from a cloud or clear air. They frequently occur in polar regions in stable air and only at very low temperatures. Ice crystals are not assigned an intensity. I was on watch at the Lovelock, Nevada, Flight Service Station (FSS) on a very cold, clear day. Ice crystals were sublimating right out of the clear air. It was a beautiful sight—but of absolutely no significance to aviation or anything else. Ice crystals generally do not result in the formation of structural icing. Ice crystals can accumulate in inlets and ducts and dislodge as a mass that can create choking and possible engine damage. A mixed condition where both ice crystals and supercooled droplets exist probably constitutes the worst condition for engine and intake icing.

Ice pellets, formerly *sleet,* are grains of ice consisting of frozen raindrops or largely melted and refrozen snowflakes. They fall as continuous or intermittent precipitation. Ice pellet showers are pellets of snow encased in a thin layer of ice formed from the freezing of droplets intercepted by the pellets or water resulting from the partial melting of the pellets. Ice pellets do not bring about the formation of structural ice, except when mixed with supercooled water. Frequently, ice pellets or ice pellet showers indicate areas of freezing rain above. (On November 5, 1998, the previous international abbreviation for ice pellets was changed to *PL*. It seems this was required because in certain combinations with other weather contractions it resulted in offensive language. It is nice to know that our METAR and TAF reports are politically correct.)

Hail is precipitation in the form of balls or irregular lumps of ice, always produced by convective clouds, nearly always cumulonimbus. An individual ball is called a *hailstone*. The largest sized hailstones and greatest frequency of occurrence in a mature cell are usually between 10,000 and 30,000 ft, but encounters can occur in clear air outside a storm cell. Hail also can include other forms of frozen precipitation with differing origins. Thunderstorms that are characterized by strong updrafts, large LWCs, large cloud drop sizes, and great vertical heights are favorable to hail formation. The violent updrafts keep hailstones suspended for several up and down cycles. Each cycle adds a layer to the hailstone, until it can no longer be suspended in the cloud.

Hail can cause severe damage to objects on the ground as well as aircraft. Blunted leading edges, cracked windscreens, and frayed nerves are a common result of a hail encounter. Like most thunderstorm hazards, avoidance is the only solution.

Many encounters are of a mixed condition consisting of ice crystals, snow, and supercooled water droplets. The mixed-cloud condition is basically an unstable condition. It is extremely temperature-dependent and may change quite rapidly. This condition can be very critical from an icing standpoint because the aggregate of impacting ice crystals, snow, and water droplets can adhere rapidly and roughly

CASE STUDY

There have been a number of multiple-turbine-engine power-loss and instability occurrences, forced landings, and accidents attributed to operating airplanes in extreme rain or hail. Investigations have revealed that rain or hail concentrations can be amplified significantly through the turbine engine core at high flight speeds and low engine power conditions. Rain or hail may degrade compressor stability, combustion flameout margin, and fuel-control run-down margin. Ingestion of extreme quantities of rain or hail through the engine core ultimately may produce engine problems, including surging, power loss, and flameout. Therefore, the Airworthness Standards of 14CFR Parts 23, 25, and 33 have been changed to reflect this concern. Pilots must be familiar with this phenomenon and comply with the manufacturer's recommendations. Under extreme conditions, avoidance may be the only safe alternative.

to the airframe, causing a significant reduction in aircraft performance. The effects of mixed conditions may form the classic "severe" icing environment.

Land and water surfaces underlying the atmosphere significantly affect cloud and precipitation development. As moist air moves upslope and cools adiabatically, condensation produces clouds and precipitation. Showers occur if the air is unstable; when the air is stable, precipitation will be more widespread and steady. As the air moves over the crest and downslope, it heats adiabatically, and precipitation ceases and clouds dissipate.

Another phenomenon is *lake effect*. Often in winter, cold air moves over relatively warm lakes. The warm water adds both heat and water vapor to the air. The added heat makes the air unstable, resulting in showers to the lee, or downwind, sides of the lakes. Since it is winter, snow showers develop downwind. These snow showers can be heavy and produce severe aircraft icing. This often occurs in the Great Salt Lake area of Utah and around the Great Lakes. In November 1996, severe lake effect caused heavy snow and the closing of Cleveland's

Hopkins International Airport for days. During the period, several aircraft slid off the runway.

Like lake effect, whiteout is another winter hazard. *Whiteout* is an atmospheric optical phenomenon in which the pilot appears to be engulfed in a uniformly white glow. Neither shadows, horizon, nor clouds are discernible; sense of depth and orientation is lost. Whiteout occurs over an unbroken snow cover and beneath a uniformly overcast sky when light from the sky is about equal to that reflected from the snow surface. Blowing snow may be an additional cause.

How do cloud droplets grow? A typical cloud droplet is about 100 times smaller than a raindrop, with an average diameter of 20 μm. When a cloud is in equilibrium, the size of the droplet does not change. The condensation of water molecules on the droplet balances evaporation. However, if equilibrium does not exist, the droplet will either increase or decrease in size.

To keep cloud droplets in equilibrium, the air must be supersaturated—relative humidity must be greater than 100 percent. Therefore, removing moisture from the air causes the droplet and cloud to evaporate; increasing the moisture in the air causes the droplet to grow, eventually resulting in precipitation.

The merging of two water drops into a single, larger drop is called *coalescence.* Not every collision results in coalescence. The process depends on a number of factors, which include the relative velocity of impact and the size of the colliding drops.

Typical cloud droplet size is 20 μm, large cloud droplets are 100 μm. As droplets grow, they can no longer be suspended in the air and begin to fall as precipitation. Typical drizzle droplets are 200 μm, with raindrops varying from about 1000 to 5000 μm.

What does this have to do with aircraft icing? As we shall see, the larger the droplet, the more severe is the icing, and the more difficult it is to remove from the aircraft. The greater the vertical motion of the atmosphere, the larger the droplet that can be supported. For example, in cumulus clouds, convective updrafts of at least 1 m/s, often exceeding 10 m/s, occur. A 100-μm droplet caught in the updraft rises, collides with and captures smaller droplets, and then grows to about 1000 μm. At this point the droplet balances gravity and remains suspended until it grows larger. As the droplet falls, it grows even larger. By the time it reaches the bottom of the cloud, it may be over 5000 μm (5 mm) in diameter!

In middle and high latitudes, such as the United States, clouds often extend upward into areas where the air temperature is below freezing. These clouds are know as *cold clouds. Cold* refers to the fact that part of the cloud is above the 0°C isotherm—line of equal temperature.

In the cold air just above the 0°C isotherm, virtually all cloud droplets consist of supercooled liquid water (Fig. 1-4). Why? With small cloud droplets, less than 20 μm, the onset of ice crystal formation begins between −9 and −15°C. Where fewer but larger droplets exist, ice crystal formation begins between −4 and −8°C. The transformation of cloud particles from liquid water droplets to ice crystals is called *glaciation.* For example, the upper portion of a cumulonimbus cloud is glaciated.

Even at higher altitudes, for example, where the temperature is −10°C, only one ice crystal exists for every 1 million liquid droplets. At −20°C, water droplets still outnumber ice crystals. Ice crystals aloft can reduce the icing threat to an aircraft by scouring out supercooled cloud droplets. Not until we reach −40°C do we find only ice crystals. This is why cirrus clouds that form above 20,000 ft, where the standard temperature is less than −25°C, are almost always composed of ice crystals and do not present an icing threat.

When moist air comes in contact with a cool surface and then cools to its dew point, dew appears. If the dew-point temperature is below

DISTRIBUTION OF ICE AND WATER

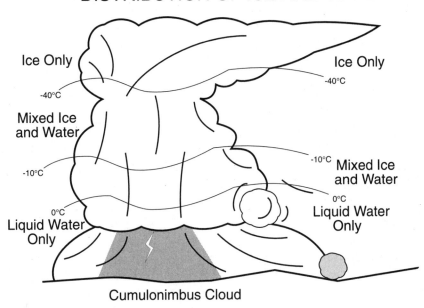

Ice Only

-40°C

Mixed Ice
and Water

-10°C

Liquid Water
Only

0°C

Ice Only

-40°C

Mixed Ice
and Water

-10°C

Liquid Water
Only

0°C

Cumulonimbus Cloud

Fig. I-4. Structural icing results from supercooled liquid water that coexists with ice crystals between the temperatures of 0 to −40°C.

freezing, the water vapor sublimates—changes from a vapor directly to a solid—producing frost. Frost may accumulate in quite heavy amounts over a clear night due to the effects of nocturnal radiation. Frost is a significant icing hazard and will be discussed in detail in subsequent chapters.

Supercooled Large Droplets

Supercooled large droplets (SLDs) are freezing drizzle (supercooled drizzle droplets, or SCDDs, 30–300 μm) or freezing rain–sized supercooled water droplets, as opposed to cloud droplets (super-cooled cloud droplets, or SCCDs), which are much smaller. These supercooled water droplets, suspended in the air, freeze almost instantly when they encounter a hard surface such as an aircraft, causing ice to accumulate. Since the smaller cloud drops tend to be deflected by the airfoil, an aircraft tends to collect more of the larger drops.

Ironically, the worst icing conditions for aircraft can occur in a cloud deck that may not contain ice itself or produce any rain or snow at the ground but that is full of supercooled water. This condition may be found in a shallow stratus deck, where an unwary pilot may expect only trace to light rime icing.

A strong relationship between the existence of SCDDs in stratiform clouds and wind shear near the cloud tops occurs with temperatures warmer than $-15°C$. At temperatures colder than $-15°C$, supercooled water is scarce, and the cloud most likely is glaciated.

CASE STUDY

On October 31, 1994, American Eagle flight 4184, an ATR72, crashed near Roselawn, Indiana. The airplane was in a holding pattern, descending to a newly assigned altitude of 8000 ft. At this point, without control input from the pilots, the airplane rolled and entered a rapid descent. The airplane was destroyed on impact with the ground. All on board died.

The National Transportation Safety Board (NTSB) determined that loss of control was the probable cause. A ridge of ice had developed behind the deice boots. The ice ridge caused uncommanded (without pilot input) upward aileron movement. Since the autopilot was in use, the pilots could not detect the loss of roll control. This caused a sudden, unexpected, aileron reversal and roll.

The NTSB issued numerous recommendations that, in hindsight, might have prevented this accident. All too often NTSB recommendations are idealistic and not feasible with today's technology. Keep in mind that airplanes have been negotiating icing conditions successfully since the 1930s. As stated in the Introduction, it is not our place to judge or criticize; our goal is knowledge and understanding.

At the time of the accident, conditions favored the development of supercooled drizzle drops within a strong warm frontal zone, cloud top temperatures between -10 and $-15°C$, weak radar reflectivity, and strong vertical wind shear.

Atmospheric conditions that result in the creation of large-droplet icing clouds are relatively rare, although research shows aircraft encounters to be higher than expected. Since this hazard may be greater than anticipated, its potential catastrophic results cannot be ignored.

More than ten years ago, research aircraft first inadvertently encountered SCDDs in what appeared to be ordinary stratiform clouds. Airframe icing that resulted was severe. In one case as little as a 5-minute exposure resulted in severe icing. SCDDs have a greater adverse effect than smaller, cloud-sized droplets. Clouds at temperatures warmer than $-10°C$ with high LWC also may result in extreme performance degradation. Analysis has shown that SCDDs were present in the holding pattern at the time of the American Eagle accident. SLD icing is an active area of interest in the icing community, and the FAA and other government organizations have placed a high priority on investigating this phenomenon.

Large droplet icing conditions are defined as icing clouds composed of water droplets with a median volumetric diameter (MVD) ranging from 40 to 400 μm. More common icing clouds consist of droplets with MVDs ranging from 10 to 40 μm. FAA aircraft certification for flight into known icing conditions is based on supercooled drops with MVDs of less than 50 μm.

Earlier research found that aircraft ice formed from large-droplet icing clouds develops (accretes) further aft on aircraft surfaces than that from the more common 10- to 40-μm-droplet icing clouds. This results primarily from the fact that large droplets impact (impinge) further aft on the airfoil surfaces than do small droplets. A significant amount of runback and secondary impact also have been observed in tests. *Secondary impact* is a term used to describe the action of unfrozen water droplets blowing off the accreted ice and then impacting further downstream. Research has found that a significant amount of ice from the large-droplet icing cloud can accrete aft of conventional ice protection equipment on an airfoil. If the ice protection equipment is activated during the icing encounter, only the ice on the protected portion of the airfoil is removed, leaving a forward facing ridge of ice aft of the protected area. When the ice protection system is not activated during the encounter and the air is at or near freezing temperature, ice will accrete until a large portion of it is blown off the leading edge by aerodynamic forces. The residual ice aft of the leading edge again has a forward facing ridge.

The National Aeronautics and Space Administration (NASA) has conducted research on the large supercooled droplet phenomenon. A Twin Otter wing section was subjected to an array of large-droplet icing conditions. NASA used an icing environment with an MVD of 160 μm and an LWC of 0.82 g/cm. These tests revealed the following: An ice ridge formed aft of the active portion of the deicer boot for every experimental test run in which ice was accreted. The location, height, and spanwise extent of the ridge varied considerably. This variability was caused by random shedding of the ice.

Large-droplet ice accumulations were found to be sensitive to changes in total temperature. Temperatures at which large-droplet icing conditions exist are believed to occur in nature between a lower limit of −15°C and an upper limit of about 0°C. At 3°C, no ice accumulation formed. All impacting water ran back to the trailing edge of the airfoil and was blown off by aerodynamic forces (the relative wind). At a temperature of 1°C, less runback occurred, and ice began to form. When ice formed, a distinct ridge developed just aft of the active portion of the deice boot. At temperatures warmer than about −1 to 1°C, the ice ridge behind the active portion of the boot had a tendency to randomly self-shed the ice buildup. However, as the total temperature was decreased below −1°C, the ice ridge appeared to harden and become more resistant to self-shedding. At temperatures below −1°C, substantial ice ridges developed aft of the active portion of the boot. This can be seen in Fig. 1-5. Note

Fig. I-5. Supercooled large droplets can produce ice well aft of the aircraft's ice-protected area. *(Photo courtesy of NASA.)*

the ice buildup well aft of the ice-protected (deice boot) area on the high-pressure surface—bottom of the wing. The largest ice ridges occurred at temperatures around $-10°C$. Ice ridges heights of over 1 in. occurred on both pressure surfaces at these temperatures. Both the high-pressure and the low-pressure surfaces—top of the wing—were affected. As the temperature was varied, the ice ridge reached a relative maximum at a total temperature of $-3°C$ for 110 knots and of $-1°C$ for 170 knots.

An increase in droplet size moved the impact limits further aft on the airfoil. In addition, runback and secondary impact accumulated ice aft of the impact limits of the ice-protected area on the airfoil.

Increasing the angle of attack (AOA) caused more ice to accumulate on the high-pressure surface and less ice on the low-pressure surface. Tests were run for AOAs between $-2°C$ and about $4°C$. As the AOA increased, the extent of ice aft of the boot on the high-pressure surfaces increased significantly. As the flap setting was increased, the extent and amount of ice accumulation on the upper wing surface decreased.

Boot cycle tests were done at total air temperatures between 0 and −15°C. (*Total air temperature* is the result of ambient air temperature and aerodynamic heating. Its effects are small until the aircraft reaches high subsonic speed. However, at high subsonic speeds, total air temperature can put the aircraft in icing temperature ranges even though ambient air temperature is well below typical icing limits.) At 0°C, ice formed that was relatively weak and prone to self-shedding. A ridge formed on both pressure surfaces for all boot cycles at this temperature. At −15°C, as the boot cycle interval increased, it was observed that the amount of residual ice left on the boot increased. Boot cycle had a definite effect on the amount of residual ice left on the boot. Variation in the boot cycling time, however, did not appear to have a significant effect on the residual ice accumulation. Changing the boot-cycle interval time did not prevent the formation of an ice ridge.

NASA says the Twin Otter tests leave a great deal of work to be done. The tests produced a relatively small amount of data. Additional data are needed to generate a realistic large-droplet cloud envelope for supercooled large droplets. Furthermore, additional tests need to be conducted using various flap settings.

According to the NWS's Aviation Weather Center:

The worst icing occurs when aircraft encounter supercooled liquid water in the drizzle drop size. Obvious weather conditions are temperatures below 0°C and environments with high relative humidity. However, in order for these drizzle drops to form, the atmosphere must be undergoing upward vertical motion at slightly faster speeds than the large-scale lift that forms large cloud masses (about 1 cm/s). Too much upward vertical motion and water drops larger than drizzle size quickly develop. The optimum vertical motion is on the order of 10 cm/s. One likely cause of vertical motion of this magnitude is the slowing down of convective bubbles as they encounter decelerating forces. This condition occurs frequently at cloud tops and is a good explanation of why significant icing is often observed there.

The patterns of the conditions for significant icing (temperature, relative humidity, and slight convective potential) are numerous and quite complex. They do not lend themselves to typical input/output prediction schemes. With better analytic tools and an increased awareness of the threat of ice to aircraft safety, the AWC is better prepared to...forecast hazardous icing conditions.

The FAA recommends the following actions when SLD conditions are encountered:

1. Disengage the autopilot. Hand-fly the airplane. The autopilot may mask important cues or may self-disconnect and present unusual attitudes or control conditions.

2. Advise ATC and promptly exit the conditions, using control inputs that are as smooth and small as possible.

3. Change heading, altitude, or both to find an area that is warmer than freezing, substantially colder than the current ambient temperature, or clear of clouds.

4. When severe icing conditions exist, reporting may assist other crews in maintaining vigilance. It is important not to understate the conditions or effects of the icing observed.

Should an uncommanded roll occur, the FAA recommends the following action:

1. Reduce the angle of attack by increasing airspeed or extending wing flaps to first setting, if at or below the flaps extend speed. If in a turn, roll wings level.

2. Set appropriate power, and monitor the airspeed and angle of attack. A controlled descent is a better alternative than an uncontrolled descent.

3. If flaps are extended, do not retract them unless it can be determined that the upper surface of the airfoil is clear of ice,

because retracting the flaps will increase the angle of attack at a given airspeed.

4. Verify that wing ice protection is functioning normally and symmetrically by visual observation. If not, follow manufacturer's recommendations.

Note that these procedures are almost opposite to those for recovery from a tailplane stall, which is covered in Chapter 2.

The Formation of Supercooled Drizzle Droplets

There are two current theories for the formation of SCDDs. First, freezing rain and drizzle result from liquid precipitation, at above-freezing temperatures, that falls through an inversion of below-freezing temperatures. Second, SCDDs form in a wind-shear environment.

Freezing rain and freezing drizzle occur when drops fall through a subfreezing layer and freeze on contact with a surface whose temperature is at or below freezing. Typically, freezing rain or drizzle begins as snow, falls through a warmer layer, and melts. The melting snow may produce a bright band on Doppler weather radar. As the liquid precipitation falls through freezing temperatures, the drops become freezing rain, freezing drizzle, or ice pellets. Close to the surface, a shallow, convective boundary layer may be present. Cloud-top temperature generally must be colder than $-15°C$ for significant ice to develop in the upper cloud. At the surface, precipitation in the form of warm rain becomes freezing rain or ice pellets and finally snow. The typical area of SCDDs aloft is where freezing rain or ice pellets occur at the surface.

Supercooled drizzle droplets may form in a wind-shear environment. Drizzle drops may originate directly from a coalescence process acting on LWC in cloud and drizzle drops. Cloud-top temperatures should be warmer than about $-15°C$ for SCDDs to occur. Should temperatures be colder, the cloud likely will be glaciated. Wind shear appears to either initiate or accelerate the coalescence process. SCDDs may be encountered near the top of the cloud, without the occurrence of freezing drizzle at lower levels. This is due to either insufficient time or LWC for droplet growth to become drizzle size or evaporation of the drops between the cloud and the ground. Should a strong inversion be present above the cloud top, it is possible for SCDDs to exist near the top of the cloud and warm drizzle to fall below the freezing level.

A summary of recommendations the NTSB made following the Roselawn accident:

1. Revise existing aircraft icing intensity reporting criteria by including nomenclature that is related to specific types of aircraft. The FAA, along with the NWS, continues to meet and confer on this subject. In April 1997, the FAA published its *Aircraft Icing Plan*. This plan addressed this recommendation. This program is progressing with all due bureaucratic speed.

2. Publish the definition of the phrase *icing in precipitation* in appropriate aeronautical publications, emphasizing that the condition may exist both near the ground and at altitude. The FAA again responds with its *Aircraft Icing Plan*. (The phrase *icing in precipitation* will be discussed in Chap. 4.)

3. Revise icing certification testing regulations to ensure that airplanes are properly tested for all conditions in which they are authorized to operate. The FAA responds that current regulations ensure that airplanes are safe for operation in icing conditions for which they are certified. Remember—and this will be emphasized again—that no aircraft is certified for flight in severe icing. The FAA acknowledges that airplanes may encounter icing conditions not defined in their certification.

The NTSB also has recommended that regulations be revised to ensure they are compatible with the published definition of severe icing and to eliminate the implied authorization of flight into severe icing conditions. And again, the FAA responds with its *Aircraft Icing Plan*. The FAA, however, has issued bulletins to direct principal operations inspectors to ensure that training programs for persons operating aircraft for hire include information about flight into freezing rain and freezing drizzle.

4. Conduct or sponsor research and development of on-board aircraft ice protection and detection systems that will detect and alert flight crews when they encounter freezing rain and freezing drizzle and airframe ice. Once more, this is addressed in the *Aircraft Icing Plan*.

Supercooled Drizzle Droplet Recognition

So far the discussion of SLDs does not specifically address small aircraft or most general aviation operations. However, what we can do is analyze what has been presented and draw some general conclusions. This was done in the *Pilot's Guide to Weather Reports, Forecasts, and Flight Planning* with microburst and low-level wind shear recognition, avoidance, and recovery.

It appears that the most likely areas for SLD occurrence are

- 25 to 300 mi ahead of a warm front

- 30 mi either side of an occluded front

- 25 to 130 mi ahead of a Pacific cold front

- 25 to 130 mi behind an Arctic front

These areas are illustrated in Fig. 1-6. One should not be surprised by these typical location for SCDDs. Twenty-five to 300 mi ahead of a warm front is the typical location of an inversion produced by the warm air overriding cooler air. The same type of temperature profile exists with frontal occlusion. Often cold air is trapped in the valleys of the Pacific Northwest. Relatively warmer air is brought in with an approaching cold front. Liquid precipitation may fall into the colder, subfreezing air in the valleys, producing freezing rain, freezing drizzle, and ice pellets. The same situation occurs behind an Arctic front, with cold, subfreezing air near the surface and relatively warmer air aloft. (In winter, Arctic fronts occasionally can approach the Gulf Coast.) Should a flight take a pilot anywhere near these weather occurrences, he or she can apply the criteria in the following paragraphs.

No mention has been made of stationary fronts. Typically, these fronts are benign, with little in the way of active weather. However, stationary fronts need to be watched. When warm, stable air from the west replaces cooler air to the east, the weather will be similar to that of a warm front with stable air. This has the potential to create SCDDs, especially if it develops into a warm front. If the air is unstable, cumulus cloud and thunderstorms may occur.

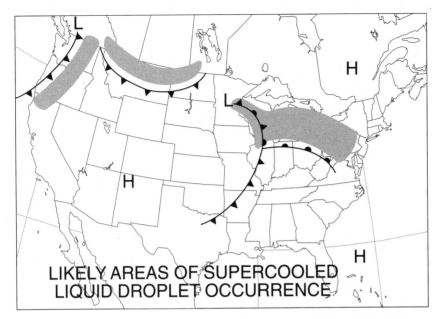

Fig. 1-6. Likely areas of supercooled large droplets are associated with an inversion, where liquid precipitation falls into subfreezing temperatures.

Look for inversions aloft, especially associated with a stable warm front. The position of warm fronts can be found on the surface analysis chart and forecast on significant weather prog charts. Along with the freezing level, inversions aloft are depicted on the freezing-level panel of the moisture/stability charts.

Cloud-top temperatures can be determined from reported or forecast tops and the wind and temperatures aloft forecast. Along with wind and temperature forecasts, AIRMET Zulu may provide temperatures at the planned flight level. Areas of high humidity are shown on the average relative humidity portion of moisture/stability charts. The most severe SCDD icing occurs between −1 and −10°C near the cloud tops. Be cautious of shallow stratus layers.

Look for reports of freezing rain, freezing drizzle, or ice pellets at the surface in METARs or PIREPs. But remember, SCDDs that form in a wind-shear environment may produce warm drizzle with no freezing precipitation at the surface. (Unfortunately, with icing there are no absolute answers—except avoidance!)

There are other indicators of an SCDD environment. However, pilots do not have direct access to these data via NWS observed or forecast charts. These are

- Bright band on Doppler weather radar

- Shallow convective boundary

- Wind-shear environment with a temperature warmer than $-15°C$

- Strong inversion above the clouds

Pilots have indirect access to these SCDD indicators through NWS forecast products, specifically AIRMET Zulu in the AIRMET Bulletin. Additional indirect access will become available through future products and developments. Both will be discussed in detail in Chap. 4. Finally, the techniques contained in Chap. 6 for dealing with and avoiding the icing hazard apply equally to supercooled large droplets as they do to the more typical, smaller supercooled droplet.

CUES FOR DETECTING SLD ICING

1. *Visible ice on the upper or lower surface of the wing, aft of the active part of the deicing boots.* It may be helpful to look for irregular or jagged lines or pieces of ice that are self-shedding. These areas need adequate illumination for night operations.

2. *The aft limit of ice accumulation on the propeller spinner.* Nonheated propeller spinners are useful devices for sorting droplets by size. SLD icing will extend beyond normal ice limits.

3. *Granular dispersed ice crystals or total translucent or opaque coverage of the unheated portions of the front or side windows.* This may be accompanied by other ice patterns on the windows such as ridges. These patterns may occur within a few seconds to $1/2$ minute after exposure to SLD conditions.

4. *Unusually extensive coverage of ice, visible ice fingers, or ice feathers on parts of the airframe not normally covered by ice.*

CUES FOR DETECTING SLD ICING (Continued)

The following are additional cues for detecting SLDs at or near-freezing temperatures:

1. *Visible rain, consisting of very large water droplets*
2. *Droplets splashing or splattering on impact with the windscreen*
3. *Water droplets or rivulets streaming on heated or unheated windows*
4. *Weather radar returns showing precipitation*

In reduced-visibility conditions, select taxi/landing lights "On" occasionally. Rain also may be detected by the sound of droplets impacting the aircraft. Droplets covered by icing certification envelops are so small that they are usually below the detectable threshold. Water droplets or rivulets are an indication of high LWC. Radar returns showing precipitation suggest that increased vigilance for all the cues is warranted. Evaluation of the radar may provide alternative routing possibilities.

CASE STUDY

To further illustrate the effects of SCDDs, let's examine the March 4, 1993 incident involving an ATR-42, smaller cousin to the ATR-72 of the Roselawn accident, while approaching Newark, New Jersey. On this day, tops ranged to only about 11,000 ft, with the freezing level at 3000 ft. The NTSB reported that "The captain described the icing conditions as light to occasionally moderate, and that there was mixed rime and clear ice. He stated that the ice accumulated to about 3 inches aft of the protected surfaces and as far aft as he could see. The ice was observed to be only on top of the wing. In addition, there was moderate turbulence in the icing environment, including wind shear of about 10 knots." The autopilot disconnected and the airplane rolled uncommanded about 50°C to the right. In this case the crew recovered and was able to land safely.

This incident occurred more than a year before the Roselawn accident. Little had been published for the pilot community about the effects of SCDDs. This incident is included for two reasons. First, in hindsight, the pilot had a number of clues: an innocent looking stratus layer, moderate turbulence, wind shear, light to occasional moderate mixed icing, and ice accumulation aft of the ice-protected area. Second, it points out how hazardous icing can be masked by a seemingly benign environment.

ICING MYTH

Stratus clouds only contain light rime icing.

It is easy to criticize the FAA, especially in hindsight. Some have made a career of it. God knows the FAA has many faults. Let's all keep in mind that there is a delicate balance between safety and overregulation. Ironically, many of those who hammer the FAA about safety are the first to holler foul at what they perceive as overregulation.

CHAPTER 2

Icing Types and Intensities

Icing affects airframes, engines, propellers and rotors, and aircraft flight instruments and radios. Icing also can occur with water droplets slightly above 0°C if the airframe temperature is at or below 0°C, such as on an aircraft descending from an area of cold air. Aerodynamic cooling also can lower the temperature of an airfoil to 0°C even though the ambient temperature is a few degrees warmer. Wind tunnel experiments reveal that saturated air flowing over a stationary object may form ice on the object when the air temperature is as high as 4°C. The object's temperature cools by evaporation and pressure changes in the moving air current. Typically, when the temperature reaches −40°C or less, it is generally too cold for airframe ice to form. Therefore, a pilot must deal with structural icing in an environment of visible moisture with temperatures between +4 and −40°C. However, icing intensities will vary widely within this temperature range.

Ice on an airfoil disrupts the smooth flow of air; it decreases lift and increases drag. Lift may be decreased by as much as 50 percent and drag may be increased by as much as 35 percent. Ice increases the weight of the aircraft and may affect the engine, reducing power output or in extreme cases causing

FACT

Ice typically forms first on small radii objects, such as pitot tubes, outside air temperature probes, and antenna masts. Pilots can use these objects to detect the beginning of an icing encounter.

engine failure. NASA testing indicated that "in the case of the commercial transport airfoil, it was found that exposure to glaze ice conditions for as little as 2 minutes can have a noticeable effect on lift coefficient, particularly at the higher airfoil angel of attack. The observed angle of stall was also affected by the presence of glaze ice. Rime and mixed ice also reduced lift coefficient and the observed angle of stall for the airfoil, although less significantly than glaze ice. Similar trends were observed for the business jet airfoil." It stands to reason that similar reductions in lift will occur with light-airplane general aviation airfoils.

Ice forms on propellers and rotors. Even a small amount of ice, if not evenly distributed, can cause stress on the piston engine mounts, propeller, and rotors. When the propeller sheds ice, a momentary increase in vibration and stress occurs. This can be very exciting as chunks of ice hit the fuselage. Ice also can form in jet engine inlets. Finally, ice affects aircraft intakes, induction systems, and carburetors.

Ice also can affect aircraft instruments. Iced-over pitot-static instruments can cause false readings or render flight instruments useless. Ice also affects radio communications by reducing antenna efficiency or causing antennas to brake off.

Ice can obstruct a pilot's view, especially on aircraft without ice protection equipment. Needless to say, it is extremely dangerous to attempt to land with an iced-over windscreen. Add to this the fact that ice can disrupt the function of control surfaces, reduce the effectiveness of brakes, and interfere with landing gear operation, and the problem magnifies.

Ice adversely affects an aircraft by

- Increasing drag
- Producing loss of lift
- Increasing weight
- Reducing or causing a full loss of power

- Interfering with control surfaces

- Reducing the effectiveness of brakes

- Interfering with landing gear operation

- Increasing vibration and structural stress

- Reducing or precluding forward vision

- Causing loss of or false instrument readings

- Causing loss of or reduced radio navigation and communications

Icing can form as slowly as $1/2$ in. per hour or as rapidly as 1 in. per minute! Icing potential exists any time visible moisture exists—clouds or precipitation. This contradicts the notion that icing can only occur in clouds. Chapter 1 mentioned how freezing rain can be the most serious icing hazard and discussed the icing potential of wet snow. Both phenomena can affect the VFR pilot. The greatest icing potential occurs between the freezing level and -10 to $-15°$C, or within a layer approximately 5000 and 7500 ft deep. Although rare, icing has been encountered in convective clouds at altitudes of 30,000 to 40,000 ft in temperatures less than $-40°$C. As mentioned previously, icing potential is enhanced by upslope from mountain ranges and lake effect.

Ice attacks three aircraft areas: aircraft structure, induction/carburetor systems, and flight and engine instruments. As we shall see, a fourth area, ground icing, is potentially hazardous to aircraft structures, induction systems, and instruments.

With a solid understanding of how ice forms, we move on to icing types: rime, clear, and mixed. This is followed by a discussion of icing intensities. The final section of this chapter deals with ice protection equipment. Manufacturers have a choice of pneumatic boot, electrothermal, fluid, electroimpluse, and hot air systems. Each system has advantages and limitations for various applications. The section concludes a discussion of the TKS ice protection system for the Commander 114.

Structural Icing

Structural icing accidents account for only about 40 percent of total accidents involving icing. Most structural icing accidents occur when a pilot continues flight into known icing, severe weather, or deteriorating weather conditions. A lesser amount occurs on approach or landing in icing conditions or with ice accumulation.

Structural icing occurs when supercooled cloud or precipitation droplets come in contact with a surface that is below freezing. The droplets form as a liquid in above-freezing temperatures and then are lifted or fall into air that is below freezing. Droplets are lifted by one of the vertical motion producers discussed in Chapter 1 or fall as liquid precipitation into a below-freezing environment (freezing drizzle and rain or ice pellets). Water droplets that form in below-freezing temperatures are already frozen and do not produce structural icing.

As well as wing and fuselage icing, tailplane or empennage stall is another hazard associated with structural icing. Over the years, some 16 commuter and transport airplane accidents have been blamed on this phenomenon. Most likely such icing has caused many more accidents, but the evidence evaporates—pardon the pun—at the crash site. Since empannage leading edges have smaller radii than the wing, ice usually forms there first. Typically, the tailplane builds ice at two to three times the rate of the wing. A tailplane stall occurs, as with the wing, when the critical angle of attack is exceeded. Since the horizontal stabilizer counters the natural nose-down tendency caused by the wing's center of lift being behind the center of gravity, the airplane will react by pitching down, sometimes uncontrollably, when the tailplane stalls.

Perhaps the most important characteristic of a tailplane stall is the relatively high airspeed at the onset and, if it occurs, the suddenness and magnitude of the nose-down pitch. Application of flaps can aggravate or initiate a tailplane stall. A stall is more likely to occur when the flaps are approaching the fully extended position, after nose-down pitch and airspeed changes following flap extension, or during flight through gusty winds. If there is the possibility of icing on the tailplane, a pilot should use caution when applying flaps during an approach and avoid slips. It may be necessary to select a runway with minimum crosswind but long enough to avoid a go-around under these conditions.

The following are symptoms of a tailplane stall:

- Elevator control pulsing, oscillations, or vibrations
- Abnormal nose-down trim change
- Any other unusual or abnormal pitch anomalies
- Reduction or loss of elevator effectiveness
- Sudden change in elevator force
- Sudden, uncommanded nose-down pitch

Recovery is opposite that of a wing stall and nearly opposite from recovery from a roll upset, as described in Chapter 1. The following are FAA recommendations, which emphasize that pilots must observe the manufacturer's recommendations regarding technique and power setting.

1. Immediately retract the flaps to the previous setting and apply appropriate nose-up elevator pressure.

2. Increase airspeed appropriately for the reduced flap extension setting.

3. Apply sufficient power for aircraft configuration and conditions.

4. Make nose-down pitch changes slowly, even in gusting conditions, if circumstances allow.

5. If a pneumatic deicing system is used, operate the system several times in an attempt to clear the tailplane of ice.

RECOMMENDATIONS

Once a tailplane stall is encountered, the stall conditions tend to worsen with increased airspeed and possibly may worsen with increased power settings with the same flap setting. Airspeed, at any flap setting, in excess of the airplane manufacturer's recommendations for the flight and environment conditions, accompanied by uncleared ice contaminating the tailplane, may result in a tailplane stall and uncommanded pitch down from which recovery many not be possible. A tailplane stall may occur at speeds less than V_{FE}.

Unlike recovery from a wing stall, instead of pitching down, the pilot must pitch up to decrease the angle of attack on the tailplane. Since this is contrary to training and may not be anticipated by the pilot, it can be extremely dangerous. Should a tailplane stall occur on approach at low altitudes, recovery may not be possible. A number of air carrier accidents have been attributed to tailplane icing and stall, and this phenomenon also can occur in light, general aviation airplanes.

Another type of icing that cannot be ignored is frost. Recall that condensation is the change of state from water vapor to liquid water. When moist air comes in contact with a cool surface and then cools to its dew point, dew appears. If the dew-point temperature is below freezing, the water vapor sublimates—changes from a vapor to a solid—producing frost.

CASE STUDY

I made a trip from Mammoth Lakes to Livermore, California, in late June. As is my habit, I planned an early-morning departure, assuming that density altitude was the most significant factor. Arriving at the airport around dawn, to my surprise, I found frost on the wings! I pointed the airplane into the sun, and in about 15 minutes I was able to wipe the melting frost from the ship. This just goes to prove that you have to be ready for, and aware of, everything.

The effects of frost may be more subtle than those of inflight structural ice. Although the airfoil's aerodynamic contour remains relatively unchanged, considerable roughness, resulting in increased drag, occurs. Under no circumstances should a takeoff be attempted with frost on the aircraft. A significant coating of frost can cause a 5 to 10 percent increase in stall speed. Just as there is no such thing as a little pregnant, there is not such thing as a little frost.

Induction/Carburetor Icing

Most icing accidents are attributed to carburetor or induction system icing. And most involved the lack, or improper use, of carburetor heat.

> **CASE STUDY**
> On a flight from Van Nuys, California, to San Francisco in a Cessna 172, I encountered light icing after an ATC instruction to climb. I periodically applied carburetor heat. Something unusual occurred. With carburetor heat on, the engine ran fine; with it off, the engine faltered. On the ramp at San Francisco I parked next to a Navion that also had flown from the Los Angeles Basin but at a higher altitude, encountering more ice. Sure enough, in the Navion's air filter was a large chunk of ice. I realized that the carburetor heat in the Cessna 172 was functioning as an alternate air source. I am sure that this seems ridiculously obvious; it didn't at the time, which illustrates the hazards of learning by experience.

The induction system includes the air filter, ducting, and fuel metering device. Induction system icing consists of any ice accumulation that blocks any component of the system.

Air filter icing occurs when flying in areas of visible moisture with temperatures at or below freezing—the same conditions that produce structural icing. For VFR pilots, air filter icing should occur only when flying in areas of freezing precipitation or wet snow.

Induction system icing takes place anytime structural icing occurs. A symptom of air filter icing is a more or less gradual decrease in power. Should air filter icing occur, apply carburetor (carb) heat or alternate air. (Some aircraft are equipped with automatic alternate air. Know and follow the manufacturer's recommendations.) The application of carb heat or alternate air bypasses the air filter. Leave carb heat or alternate air on until above-freezing temperatures melt the ice. The use of carb heat or alternate air results in unfiltered air entering the induction system. Except for operational checks, avoid engaging either control on the ground.

In addition to air intake icing, normally aspirated engines can develop ice in the carburetor throat. The vaporization of fuel, along with the adiabatic expansion of air as it passes through the fuel discharge nozzle, venturi, throttle valve, and passages to the engine, causes sudden and significant cooling (Fig. 2-1). If the air temperature drops below the dew point, water vapor in the air condenses into water droplets. Therefore, water can form in the carburetor with cloudless skies. This cooling can reduce the temperature in the carburetor to below freezing, and, with sufficient moisture present, ice forms. Known as *carburetor icing,* ice can form with outside air temperatures as high as 32°C. The formation of carburetor ice restricts engine power and may result in complete engine stoppage.

As air accelerates through the carburetor and fuel evaporates, temperatures can be lowered as much as 34°C. Whether ice will develop depends on the velocity of the fuel-air mixture, outside air temperature, humidity, and carburetor system. Conditions most favorable for carburetor ice are outside temperatures between −10 and 25°C, high relative humidity, and low power settings.

Carburetor heat preheats the air before it reaches the carburetor. Carburetor heat is usually adequate to prevent icing but may not always clear ice that has already formed. Pilots should monitor engine performance for the first signs of icing, especially during

CARBURETOR

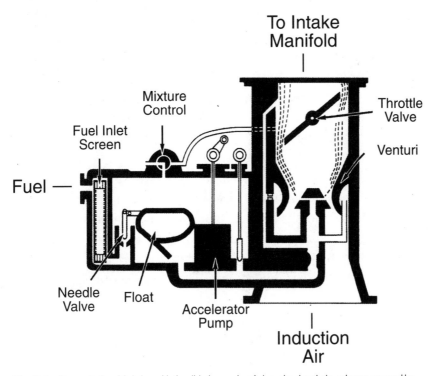

Fig. 2-1. The vaporization of fuel along with the adiabatic expansion of air passing through the carburetor causes sudden and significant cooling.

favorable conditions. When ice is detected or suspected, immediately apply full carburetor heat. Leave the heat on until all ice has been removed.

Loss of engine rpm's is the first indication of carburetor icing on airplanes equipped with a fixed-pitch propeller. On airplanes with a constant-speed propeller, carburetor icing results in a drop in manifold pressure. Engine roughness may accompany these symptoms, along with a drop in exhaust gas temperature.

Using full carburetor heat initially will cause an additional loss of power and engine roughness. The added loss of power and roughness result from the richer mixture due to warmer, less dense air, and melting ice passing through the engine. You must resist the natural urge to remove carb heat. Leave full heat on until an increase in rpm's or manifold pressure and smooth engine operation resume.

Under severe conditions it may be necessary to leave carburetor heat on for an extended period. When carb heat must be left on, relean the mixture for maximum rpm's and smoothest operation. A cruise power setting of 75 percent or less with any amount of heat will not damage the engine.

On a flight from Page, Arizona, to Las Vegas, Nevada, in a Cessna 150, conditions for carburetor ice were ideal. The temperature was about 15°C in rain showers. At the first indication of carb ice, heat was applied. It cleared up the ice, but as soon as it was removed, the rpm's began dropping. The only solution was to leave carb heat on and live with the hundred or so loss in engine rpm's, even after releaning the mixture.

Carbureted engines are more susceptible to icing during reduced-power operation. Some aircraft and engine manufacturers recommend the use of carburetor heat during all power reductions; others, only when ice is suspected. Pilots should know and follow the aircraft manufacturer's recommendations. If full power is required, such as a go-around, full carburetor heat and full power may cause early detonation or engine damage. It certainly will prevent the engine from developing full power, which might be critical in low-power aircraft at high-density altitudes. Again, know and follow the manufacturer's recommendations.

Some airplanes have a carburetor air temperature gauge. This gauge measures the temperature in the carburetor throat, near the throttle valve. The yellow arc indicates temperatures at which icing is most likely to occur. This yellow arc typically ranges between −15 and 5°C. If the air temperature and moisture content of the air make carburetor icing improbable, the engine can be operated within the

CASE STUDY

Pay attention! I had remained overnight in Amarillo, Texas, because of a line of thunderstorms that approached from the west. The Cessna 150 was parked into the wind when torrential rains moved through the area. The next morning was clear, with abundant surface moisture, temperature was about 15°C, and there was nearly 100 percent relative humidity.

My first clue of trouble was the increased throttle setting required to obtain idle rpm's. Engine runup also took more throttle than usual. I suspected carburetor ice and a water-saturated air filter because of the conditions. I had 13,000 ft of runway.

Full throttle only gave me about 2200 rpm's. The increased ground run to rotation speed—about 7000 ft—should have been another clue. I was off the ground with no runway remaining and at 200 ft of altitude when the engine began losing rpm's. I applied carburetor heat, and the engine began running very rough, producing about 1700 rpm's.

There was a tremendous psychological urge to reduce carb heat and get that rpm back. I was preparing to crash straight ahead, but the engine was still producing power, and I decided to make a 180° turn and land on a taxiway. Then I informed a rather surprised tower controller of what happened; remember, a pilot's first job is to fly the airplane.

After running the engine for 20 minutes, and one aborted takeoff later, I launched into the air. The engine performed normally above the shallow, moist layer. This is a perfect example of having the clues and ignoring them. I was extremely lucky.

yellow arc. When ice is suspected or detected, the pilot can apply enough heat to raise the temperature out of the danger zone. Some carburetor air temperature gauges have a red radial line that indicates the maximum recommended carburetor inlet temperature. Additionally, a green arc may be incorporated to indicate normal operating range. Do not use partial carburetor heat without a carburetor air temperature gauge. Applying partial heat or leaving it on for an insufficient time may aggravate the situation.

The following should alert pilots to the potential of carburetor icing:

• Temperatures between −10 to 25°C

• High relative humidity

• Low power settings

Note also that carburetor icing can occur at temperatures as high as 32°C.

Ideal temperatures for carburetor ice occur in most parts of the country, more in northern areas than in southern and greater in winter than in summer. However, many pilots have been lulled into a false sense of security during summer months with relatively high temperatures and high humidity. This typically occurs in the Midwest and East during the summer season. Follow the manufacturer's recommendations—most require the use of carburetor heat when the rpm/manifold pressure is out of the green arc.

Figure 2-2 shows a composite carburetor icing probability chart. This chart was derived from FAA and Transport Canada carburetor icing probability charts. Like structural icing, carburetor icing is complex, and there are few hard and fast rules. The probability chart graphically depicts carburetor icing potential.

Instrument Icing

The pitot-static system consists of the pitot tube, pitot line, and static ports or vents. (The pitot tube is named after its inventor Henri Pitot, an eighteenth-century French hydraulic engineer.) The pitot-static system is illustrated in Fig. 2-3. The pitot tube provides ram air to the airspeed indicator through the pitot line. Static ports vent the airspeed indicator, altimeter, and vertical speed indicator to outside, or ambient, air.

Some pitot tubes or pitot masts contain an electrical heating element. A switch on the instrument panel activates pitot heat. Like

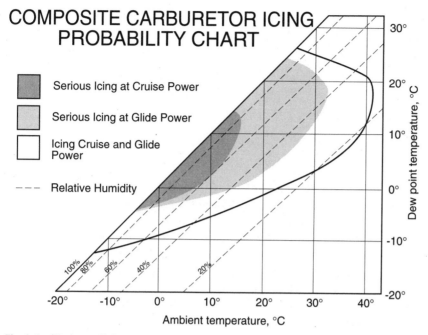

Fig. 2-2. This chart graphically depicts carburetor icing potential, but like structural icing, carburetor icing is complex, and there are few hard and fast rules.

PITOT-STATIC SYSTEM

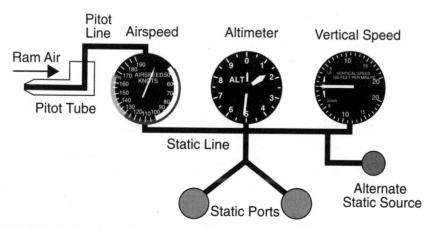

Fig. 2-3. The pitot-static system provides ram and static air to power the aircraft's pressure-driven flight instruments.

carburetor heat, pitot heat prevents ice from forming on and blocking the pitot tube. (This system is typical of Cessna and Beechcraft airplanes, when installed.) VFR pilots can expect pitot icing when flying in freezing precipitation or wet snow. An iced-over or blocked pitot results in loss of correct airspeed indications.

CASE STUDY

As a young, new instrument flight instructor, I was with a student on a flight from Van Nuys to Lancaster's Fox Field in California. En route, we were just in the cloud bases, just above the freezing level. Sure enough, the pitot tube iced up. The rime ice formed a perfect cone and grew about three-quarters of an inch into the airstream. I very nonchalantly pointed this out to my student. Then I suggested that we turn on the pitot heat. The ice disappeared in seconds. As we shall see, pitot heat is an anti-icing device and should be turned on prior to entering possible icing. Here again, for me, the test came before the lesson!

These days, Cessna aircraft are placarded with "Pitot heat must be on when operating below 40°F in instrument meteorological conditions." There is no question that pitot heat should be on when operating in clouds or precipitation with temperatures at or below about 2°C. I suppose that 40°F is a nice round, conservative number.

The airspeed indicator is the only instrument affected by a blocked pitot tube. Should this occur, the pilot can control the airplane using power and pitch attitude. Each pilot should be familiar with appropriate power settings and attitudes for various phases of flight (climb, cruise, descent, and approach).

Static ports are normally located on the sides of the fuselage. Should the ports become blocked, all three pitot-static instruments would give erroneous readings. On some airplanes the static ports are part of the pitot mast and heated when pitot heat is activated. (This system is typical of Piper airplanes, when installed.)

Many aircraft are equipped with an alternate static source, vented inside the cabin, for emergency use. Static pressure inside the cabin is usually lower than outside static pressure. Therefore, pilots should expect the following anomalies:

- The altimeter reads higher than normal.
- Indicated airspeed is greater than normal.
- The vertical speed indicator shows a momentary climb and then operates normally.

Should the static ports become blocked without an alternate static source, the pilot has the option of breaking the glass on the vertical speed indicator. Loss of the vertical speed indicator has the least impact, and the gauge is the least expensive to replace.

Ground Icing

Less than 10 percent of icing accidents involve icy runways. Other ground icing accidents result from frost, ice, or snow left on the aircraft. If an aircraft is parked in an area of blowing snow, special attention must be paid to openings in the aircraft where snow can enter, freeze solid, and obstruct operations. These openings must be free of snow and ice before flight. Additional caution is required during taxi and takeoff to avoid water and slush being thrown onto the aircraft and freezing during the flight. More is given about these hazards in Chapter 6, "Cockpit Strategies."

Figure 2-4 shows an airplane that is covered with a thick coat of snow from an East Coast snow storm that occurred the preceding day. With cold temperatures, even under sunny skies, and without deicing or a heated hangar, aircraft can remain ice-covered for days.

I had remained overnight at Farmington, New Mexico. Under clear skies, calm winds, high humidity, and temperatures that

FACT

Instrument icing has caused jet air carrier as well as general aviation accidents. A Boeing 727 was lost because of an iced-over pitot tube. As static pressure decreased during the climb, the airspeed indicator showed speed increasing. The autopilot attempted to hold airspeed by increasing pitch, resulting in a stall. A Boeing 737 crashed because of an iced-over engine power sensor—the airplane was simply not developing takeoff power, even though the instruments indicated so.

CASE STUDY

A number of years ago, I had spent several days at Lake Tahoe, in the Sierra Nevada Mountains. It was November, and snow had fallen the previous day. When I arrived at the airport the following day, there was a high overcast, and it was cold—below freezing. There was about three-quarters of an inch of snow on the airplane. While attempting to start the Cessna 172, I overprimed the engine, and the carburetor air box caught fire—but that's another story. Based on zero experience with ground icing, I assumed the snow would blow off during the takeoff roll. Not! Fortunately, an experienced tower controller suggested we remove the snow.

Well, we had to scrape the snow off with an ice scraper! This is another perfect example of learning by experience. After reaching home, I calculated that there was about 300 lb of snow on the airplane! (On a Cessna 172, that's approximately a 15 percent increase in the maximum certified gross weight.) However, reduced aerodynamic efficiency is typically the greatest hazard from frost, ice, or snow left on an aircraft.

Fig. 2-4. In cold temperatures, even under sunny skies, without deicing or a heated hangar, aircraft can remain ice covered for days.

dipped below freezing, conditions were perfect for frost. Sure enough, frost formed on the airplane.

Without deicing, a chartered Cessna Citation jet and I were not going anywhere. The Citation had the local operator deice the airplane. I used the free fusion heater, supplemented by manual enhancement to deice the Cessna 172. I waited for the sun to come up, which partially melted the frost; then I wiped off the rest of the ice! Recall that I do have experience in this area. The bottom line: All frost, ice, and snow must be removed from an aircraft before flight!

Icing Types

Icing is classified by its formation and appearance. Below are the three standard classifications as described in the *Aeronautical Information Manual* and other government publications.

- Rime ice
- Clear ice
- Mixed ice

There have been a number of recent proposals to eliminate these classifications. Reasons vary. Some pilots have observed that most inflight icing is a combination of rime and clear ice, or mixed icing. Another opinion holds that most pilots really do not understand the classifications; therefore, they cannot correctly report icing type. I am sure that there is some foundation for each point of view. However, at least for the present, pilots should understand and use these terms when reporting icing. Therefore, let's move on and define each icing type.

RIME ICE

Rime ice appears as a milky, opaque, and granular deposit with a rough surface (Fig. 2-5). Rime ice normally forms when small, supercooled water droplets instantaneously freeze on impact with the aircraft. The instantaneous freezing traps a large amount of air, giving the ice its

Fig. 2-5. Rime ice is a milky, opaque, and granular deposit with a rough surface formed when small supercooled water droplets instantaneously freeze on impact with the aircraft. (Photo courtesy of NASA.)

opaque hue and making it very brittle. Rime ice is most frequently encountered in stratiform clouds at temperatures between 0 and −20°C. There is typically a low risk of rime icing at temperatures less than −30°C. Rime ice is lighter in weight than clear ice, and its weight is typically of less significance. Rime icing conditions usually are widespread due to the character of stratiform clouds. Because it is brittle, rime ice is relatively easy to remove with ice protection equipment.

CLEAR ICE

Clear ice appears glossy, identical to glaze (Fig. 2-6). Clear ice forms when large, supercooled water droplets flow over the aircraft's surface after impact and freeze into a relatively smooth sheet of solid ice, although it also can be rough. It is smooth when deposited from large, supercooled cloud droplets or raindrops that spread, adhere to the surface of the aircraft, and freeze slowly. If mixed with snow, ice pellets, or small hail, it is rough, irregular, and whitish, as shown in Fig. 2-6. The deposit then becomes very blunt-nosed with rough bulges building out against the airflow. Clear ice is the most serious of the various forms of ice because it adheres so firmly to the aircraft. It is encountered most frequently in cumuliform clouds or freezing precipitation. Brief but severe accumulations occur at temperatures between 0 and $-10°C$ and in cumulonimbus clouds down to about $-25°C$, with reduced intensities at lower temperatures of about -20 to $-30°C$. There is typically a low risk of clear icing at temperatures less than $-40°C$. Clear ice is usually not as widespread as rime ice because it occurs in cumuliform clouds, but its intensity tends to be more severe. Clear ice is hard, heavy, and tenacious, more difficult to remove than rime ice.

A condition favorable for rapid accumulation of clear ice is freezing rain below a frontal surface. Icing also can become severe in cumulonimbus clouds along a surface cold front or above a warm front. Icing is also more probable and more severe in mountainous regions than over flat terrain. Mountain ranges cause rapid upward vertical currents on the windward side that support large water drops. The movement of frontal systems across mountain ranges often combines frontal lift with upslope effect, creating extremely severe icing zones, with the most severe icing taking place above the crest and to the windward side of the ridges.

Fig. 2-6. Clear ice is glossy, formed when large supercooled water droplets flow over the aircraft's surface after impact and freeze. (Photo courtesy of NASA.)

MIXED ICE

A mixture of rime ice and clear ice (mixed ice) is a hard, rough, irregular, whitish conglomerate formed when supercooled water droplets vary in size or are mixed with snow, ice pellets, or small hail. Deposits become blunt, with rough bulges building out against the airflow. Of the three types of ice, mixed ice is the most difficult to remove.

Icing Intensities

Pilots, especially new or low-time pilots, have a tendency to overestimate the intensity of icing. This is particularly true for pilots flying airplanes without ice protection equipment. Why? Usually they have no experience with ice and are unable to correctly relate icing intensities. For example, a recently rated instrument pilot, after experiencing his second encounter with icing in a Cessna 172, reported the intensity as severe. The encounter lasted about 30 minutes, and the pilot was unable to maintain altitude and was forced to descend. This description, however, is only of moderate intensity! Unfortunately, such pilots then believe they have experienced severe conditions and may not heed future reports or forecasts. More about this in Chap. 3 when we discuss pilot weather reports (PIREPs) and Chap. 4, "Icing Forecasts."

In Chapter 1 we discussed the physics of icing. If you gained only one concept, it should be that icing is an extremely complex phenomenon that is difficult to generalize. As pilots, we usually do not have all the meteorologic data or knowledge to apply such factors as wind-shear environment, cloud-top temperatures, and relative humidity aloft. The rate of ice formation and ice shape are affected by many factors, including liquid water content (LWC), drop size, airspeed, temperature, and size and shape of the aircraft. Recall the closing paragraphs of Chapter 1, where the March 4, 1993, Newark, New Jersey, ATR-42 incident was discussed. The following discussion relating locations and temperatures to icing intensities addresses is only generalizations; it in no way implies that greater intensities cannot occur under specific atmospheric conditions.

Icing intensities are classified for reporting purposes in the *Aeronautical Information Manual* and *Aviation Weather Services*. However, these descriptions are vague and difficult to relate, especially for pilots flying aircraft without ice protection equipment. Thus I prefer more descriptive definitions.

Trace Icing

Trace icing describes a condition where ice becomes perceptible. It appears as a white line on aircraft leading edges. The rate of accumulation is slightly greater than the rate of sublimation. When you see this, the time is right to start planning an out. It is not hazardous, even though ice protection equipment is not available, unless it is encountered for more than 1 hour.

Your passengers may admire how pretty it looks on the wing; Air Traffic Control (ATC) has just instructed you to climb. You advise them that icing is probable and request a lower altitude. The controller calmly replies that in that case "you can declare an emergency or land." Shortly, you are handed off to the next controller. You inquire about a lower altitude, and the controller responds, "Is that Terry up there?" (a friend at Los Angeles Center) A lower altitude is approved in about 15 minutes.

On an IFR flight out of San Francisco in a Cessna 172, I had to climb through a thin stratus layer. I knew the tops were below 10,000 ft and had filed for 11,000 ft. Temperatures were between 0 and $-10°C$. There was warmer air below—well above the minimum en-route altitudes, low tops with clear skies above. I encountered a trace of rime ice. Once on top, the ice sublimated rapidly off the airplane. Uniform stratiform clouds tend to produce only trace rime icing at temperatures between 0 and $-15°C$. In stratiform clouds, about two-thirds of all icing reported occurred between -2 and $-12°C$, slightly less than one-third between -12 and $-24°C$, and the remainder below $-24°C$.

Trace icing also might be expected in areas of snow grains, since they are already mostly frozen and indicate a stable air mass, or in cumuliform clouds—except cumulonimbus—at temperatures between −25 and −40°C.

Light Icing

With *light icing,* the rate of accumulation can create a problem if the flight continues for more than 1 hour. The thin white line on leading edges begins to increase in size. Occasional use of ice protection equipment removes or prevents accumulation. If the ice protection equipment is used, ice should not present a problem. If you do not have a plan or ice protection equipment, a 180° turn strongly should be considered. Presumably, you came from an ice-free area.

My instrument student had not noticed the ice yet, my pilot friend in the back seat was hoping he had enough life insurance, and I was negotiating with ATC for a lower altitude, which they can approve in 15 miles. This will take only about 8 minutes, but each minute seems like 10. In this instance we were flying in stratocumulus clouds with temperatures of 0 to −15°C. Icing was rime. In this case, again, minimum en-route altitudes were well below the freezing level.

Light to moderate icing also could be expected initially in stable areas of snow pellets or ice pellets mixed with supercooled water or in cumuliform clouds—except cumulonimbus—at temperatures between −10 and −25°C.

Moderate Icing

Moderate icing describes a rate of accumulation which, even for short periods, becomes potentially hazardous and the use of ice protection equipment or flight course diversion becomes necessary. Ice accumulation continues, and aircraft performance decreases. If cruise airspeed decreases by about 10 knots, it is time to descend to warmer air. You can descend to warmer air, right?

On his second encounter with ice, a friend and his passengers, in an aircraft without ice protection equipment, survived moderate icing only because the terrain was lower than the freezing level. My friend had enough knowledge and training to know a descent was the only option. Aircraft control was maintained and survival accomplished. (Upon their return, I counseled this pilot. We will talk about judgment in Chapter 6.)

My friend was flying in cumuliform clouds with temperatures between 0 and −10°C, associated with a frontal zone. The type of ice was not specified; there were more important issues—like survival! Typically, rime or mixed icing occurs in these areas and may develop when flying in wet snow. For light single-, and twin-engine airplanes, a climb to the top may not be an option! Expect at least moderate icing with an inversion—warm front or overrunning warm air—as described in Chapter 1, even without the presence of supercooled large droplets.

Severe Icing

The rate of accumulation in *severe icing* is such that ice protection equipment fails to reduce or control the hazard. Immediate diversion is necessary. This is a situation where the person in the left seat very rapidly ceases being the pilot and becomes a passenger; the wing is an ice cube.

The ATR-72 accident at Lawndale, Indiana, illustrates a severe icing encounter. The aircraft's ice protection equipment failed to control the ice accumulation. Like extreme turbulence, severe icing results in an aircraft that is not controllable and can cause structural damage.

Expect the possibility of severe icing in cumulonimbus clouds, in areas above the crest and windward side of mountain ranges; with lake-effect snow, freezing rain or drizzle, hail with a large liquid content, as well as with those conditions described in Chapter 1 associated with supercooled large droplets (SLDs). Severe icing usually is localized and forms clear or mixed ice. Do not forget frost. It has the same effect as severe icing on the aircraft!

FACT

Certain pilots insist on reporting icing intensity as heavy. This is a misnomer—all ice is heavy! (It seems that the maximum intensity reported for icing was "heavy" until 1968. That's almost before my time!)

Severe icing, like extreme turbulence, is rare, transitory, and difficult to forecast accurately. No aircraft are certified for flight in severe icing. Pilots must use caution and judgment, based on their training and experience. As a popular aviation saying goes, "A superior pilot uses superior knowledge to avoid the need to employ superior skill."

Ice Protection Equipment

The FAA has issued an Airworthiness Directive (AD) requiring manufacturers to include in their approved aircraft flight manuals information for recognizing potential or existing severe icing conditions and procedures for exiting a severe icing environment. This includes the warning in the sidebar.

The following discussion applies to 14 CFR, Part 23, "Airworthiness Standards: Normal, Utility, Acrobatic, and Commuter Category Airplanes." These are major changes and requirements that have evolved over the years.

Prior to 1945, airplanes only were required to have deicer boots installed, with a positive means of deflation. There were no other references in the regulations to an ice protection system.

Ice protection was not addressed again until 1962. At that time, regulations were amended to require that information be provided to aircrews specifying the types of operation and meteorologic conditions to which an airplane was limited by its equipment. Icing was specifically addressed. The operational function of equipment also was identified. This later became know as the "Kind of Equipment List."

Regulations were again changed in 1964 and 1965. At that time, the effects of icing conditions on static pressure instruments were addressed. In 1968, the FAA instituted an extensive review of airworthiness standards. As a result, standards for engine installation ice protection were introduced.

In 1987, with the creation of the commuter category, airplanes that had weight, altitude, and temperature limitations for takeoff, en route, climb, and landing distance were being certificated. Since the operation rules preclude takeoff with ice on the airplane, the FAA determined that ice accretion on unprotected surfaces should not be a consideration until the airplane climbs through 400 ft above ground level (AGL). The FAA does not believe any significant ice will accumulate prior to 400 ft if there is no ice on the airplane at takeoff.

Not until 1991 was a requirement added for a heated pitot tube or an equivalent means of preventing malfunction due to icing. Additionally, the requirement for an alternate static air source became effective.

In 1993, the following requirements were added: a clear windscreen area necessary to ensure safe operation in icing conditions and inclusion of kinds of operations authorized in the Airplane Flight Manual (AFM).

Additionally, the pilot must be provided with necessary procedures for safe operation. This should include any preflight action necessary to minimize the potential of en-route emergencies associated with the ice protection system. The system components should be described with sufficient clarity and depth that the pilot can understand their function. Procedures should be provided to optimize operation of the airplane during penetration of icing conditions, including all flight regimes. The AFM should include procedures that advise on which condition the ice protection equipment should be activated. Emergency or abnormal procedures, including procedures to be followed when ice protection systems fail or monitor or warming alerts occur, should be provided. Information and methods also should be included to determine the remaining flight operation time for fluid systems.

Although many aircraft have limited ice protection equipment (pitot heat, prop anti-ice, alternate static source, etc.), it should never be construed as to allow flight in icing; the purpose of this equipment is emergency use only, should icing be encountered inadvertently.

This brings up the question: What is known icing? You will not find it in 14 CFR, Part 1, "Definitions and Abbreviations," or in the "Pilot/Controller Glossary" in the *Aeronautical Information Manual*. Icing is difficult to forecast and transitory in nature. Would we want a forecast of icing to forbid flight? Would a report of trace or light icing above 8500 ft, with bases at 8000 ft and tops at 9000 ft preclude flight for aircraft not certified for flight in known icing? What if the terrain was at 7800 ft and the tops were at 15,000 ft? Would we really want a hard answer? If some in the FAA regulators have their way, the definition would be any time there is visible moisture and a temperature of plus 5°C or less! Should this or a similar proposal ever be adopted, it would certainly mark the end to many useful icing PIREPs. However, if icing is reported or forecast and you fall out of the sky and survive or require emergency or special handling from ATC, you are a candidate for a violation.

FACT

A National Transportation Safety Board (NTSB) decision in 1993 held a pilot in violation of federal aviation regulations. The NTSB found that a pilot cannot pick and choose between forecasts and PIREPs. A forecast for icing is sufficient to warrant the violation.

More often than not, however, ATC is so busy and so happy to get us out of their hair that we will never hear another word. No one should interpret this as meaning that I, the FAA, or ATC condones such actions. Icing for aircraft not certified for flight in icing conditions is to be avoided! At present, the decision as to whether the flight can be made safely rests solely with the pilot, which is where it will stay until pilots prove that they are not worthy of the responsibility.

This notion of pilot responsibility relates directly to the FAA's perceived "tombstone mentality." There is no question that after an accident, politicians and bureaucrats feed on the assumption that if a regulation were in effect prohibiting the operation, the accident would not have occurred. Classic examples are the mandating of emergency locator transmitters (ELTs), 14 CFR, Part 103, "Ultralight Vehicles," and the proposed regulations after the Jessica Dubroff tragedy—which we will explore in Chapter 5.

ELTs were mandated by Congress after the crash of one of its members, the assumption being that if an ELT was on board, the congressman may not have perished. I had a conversation with a number of ultralight pilots at the 1982 Merced, California, fly-in. Their operations were unregulated at the time, and they wanted to keep it that way. I told them that as long as they did not present a nuisance to others or kill themselves or people on the ground—conduct safe operations—they would remain unregulated. Well, they didn't, and they are! After the Dubroff accident, politicians cried for regulations mandating the minimum age for pilot training. Then FAA Administrator David Hinson sent a letter to all flight instructors emphasizing their role and responsibilities in flight training. A ray of sanity.

The point: If we conduct safe operations and responsibly exercise our pilot-in-command prerogative, we can avoid overregulation.

The design of ice protection systems depends on a knowledge of icing characteristics. Systems are divided into two general types: anti-icing and deicing.

An *anti-icing system* prevents the buildup of ice and must be activated prior to an icing encounter, before entering icing conditions—visible moisture, temperature +4°C or lower. If too much ice has accumulated, the system will be required to deice the surface. This may result in runback of liquid water that may refreeze aft of the ice-protected area. These systems must be tested prior to flight into icing conditions.

A *deicing system* removes or prevents additional accumulation of ice once ice has been encountered. Improper use of the system can result in an uncontrollable accretion of ice. The system should be turned on before large amounts of ice accumulate, or some undesirable secondary effects may occur. For example, if operation of the propeller deicing system is delayed too long, the ice thrown from the propeller would be thicker than usual and could cause excessive fuselage skin damage. Like anti-icing, deicing systems must be tested prior to flight into icing conditions.

Pneumatic Boot Deicing Systems

Pneumatic boots have been the standard ice protection system for piston-engine aircraft since the 1930s. BFGoodrich Aerospace traces its lineage back over 60 years to the first commercial deicing system installed on the Northrop Alpha mail plane and then on the Douglas DC-3. Pneumatic boots are easily retrofitted, require very little power, are relatively lightweight, and are of reasonable cost.

Pneumatic boot surface deicing systems remove ice accumulations mechanically by alternately inflating and deflating tubes in a boot covering the surface to be deiced (leading edges of wings and tail surfaces). This type system is shown in Fig. 2-7 on a Cessna Citation. After boot activation, aerodynamic or centrifugal forces—when used on propellers—then remove the ice. This method is designed to remove ice after it has accumulated rather than to prevent its accumulation. Pneumatic boots are not and cannot be used for anti-icing.

In addition to the boots, primary components of a pneumatic boot system are a regulated pressure source, a vacuum source, and an air

Fig. 2-7. Pneumatic boot surface deicing systems remove ice accumulations mechanically by alternately inflating and deflating tubes in a boot.

distribution system. Miscellaneous components include a solenoid, check and relief valves, air filters, control switches and timer, and electrical interfaces including fuses and circuit breakers. The vacuum source is essential to ensure positive deflation and keep the tubes collapsed during nonicing flight conditions to minimize loss of aerodynamic efficiency. Air pumps typically multiply atmospheric pressure by a specific value to deliver sufficient pressure at cruising altitudes.

Pneumatic boot systems have limited or no application to the following aircraft components:

- Windscreens
- Engine inlets and components
- Turbofan components
- Propellers, spinners, and nose cones
- Helicopter rotors and hubs

Notwithstanding the preceding, an experimental pneumatic boot deicing system has been tested successfully on helicopter rotor blades. Ice protection of some other components, such as radomes, with pneumatic boot deicing systems has been found feasible.

Generally, a nominal thickness of ice, about $1/2$ in., is allowed to accumulate before system activation. Ice-phobic liquids may be sprayed on the boots prior to flight when icing is likely. These sprays reduce the adhesion of ice to the boot surface, improving deicing. The liquid, however, erodes away and, after a few deicing cycles or after a few hours of flight, loses most of its effectiveness.

Some aerodynamic drag penalty occurs with pneumatic boot deicing systems, but this can be minimized by recessing the surface leading edge to offset boot thickness. These systems have been in use for many years, and their repair, inspection, maintenance, and replacement are well understood. However, boot material deteriorates with time, and periodic inspection is required to determine the need for replacement.

A certain amount of pilot skill is required for safe and effective operation. Actuation when accumulated ice is too thin may result in *bridging,* where the ice formation over the boot is not cracked by boot inflation. Thus attention is required to judge whether the cycle time continues to be correct for icing conditions. This increases pilot work load, especially in darkness, since the ice accumulation rate and severity are more difficult to determine.

Over the last several years, BFGoodrich has been developing a new impulse-type mechanical ice-removal system. The system is considerably different from other dynamic systems in that a pneumatic, rather than an electrical, impulse removes surface ice.

The BFGoodrich advanced-type pneumatic deicer relies only on distorting the surface to debond ice, with a rapid movement to "launch" the ice off the aircraft. High-pressure air, either from a small compressor or tapped from an existing high-pressure source onboard the aircraft, provides the impulse. Ice is debonded at the ice–aircraft surface interface. To remove the ice, a low-deflection system is necessary. This is accomplished by imparting a sufficient amount of momentum to the ice by a rapid outward movement of the surface. This overcomes any residual adhesive forces and removes the ice.

Features include

- Removal of thin as well as thick ice accretions
- Low power consumption
- Relatively low weight
- Low aerodynamic distortion
- Long life cycle
- Composite compatible
- Automatic operation using ice detectors

The 1998 Piper Malibu Mirage is being equipped with the BFGoodrich SmartBoot ice detection and protection system. The system detects the formation and buildup of ice on wings and empannage. SmartBoot uses existing deicer boots along with an electrical conductor strip to detect ice formation. Equipment in the cabin alerts the pilot when icing conditions are encountered, indicates when to activate the system, and then verifies proper activation.

Electrothermal Systems

All thermal ice protection systems operate on the same principle: Heat is applied to an area to eliminate or prevent the formation of ice. The heat evaporates or prevents supercooled cloud droplet from freezing or melts existing ice. BFGoodrich has developed an electrothermal system for the rotor blades of the Army Apache AH-64 helicopter. Equipment for electrothermal systems is a source of electrically generated heat, a distribution system, and a control system.

In addition to airfoils and propellers, the following surfaces may be afforded electrothermal ice protection:

- Windscreen

- Essential instruments

- Engine air inlets

- Auxiliary air inlets

- Antennas

- Balance horns

Because of electrical interference from the heating element, electrothermal systems are not applicable to radomes. Of these areas, the windscreen is always given anti-ice protection, while the remaining areas my be given anti-ice or deice protection, depending on the power available and the effects of ice on the component.

Electrothermal anti-icing systems use electric heaters to maintain the temperature of the surface to be protected above freezing throughout an icing encounter. For anti-icing, the heat source must remain on throughout the icing encounter. Therefore, the areas generally anti-iced electrically are the windscreen, small air inlets, instrument probes (pitot heads), leading edges forward of engine inlets, data probes, and areas remote from any hot-air source.

Anti-icing protection is usually provided for forward-facing windscreens. The most widely used system consists of an electric current passed through a transparent conductive film or resistance that is part of the laminated windscreen. The heat also prevents internal fogging for most configurations.

In a deicing system, ice is allowed to accumulate on the surfaces to be protected and removed periodically. Electrothermal deicing systems function by rapidly applying sufficient heat to the surface interface to melt the layer of ice. Aerodynamic or centrifugal forces then remove the bulk of the ice. For efficient deicing protection, the correct amount of heat must be supplied where and when needed. If too little heat is applied, the ice may not shed, or large chunks may be shed, creating unbalanced propellers or rotors. Too much heat results in undesirable amounts of runback ice. This occurs when local heating of accumulated ice melts, and water runs back to unheated areas and refreezes.

Propeller and spinner ice protection is employed for three reasons:

1. Leading-edge ice formations may cause loss of propeller efficiency.

2. Unsymmetrical shedding of ice may result in an unbalanced propeller.

3. Large pieces of ice shed from the propeller or spinner may be ingested by the engine on turboprop aircraft.

Electrothermal deicing may be installed on the external surface of a propeller without appreciably affecting performance. Coverage is approximately 15 percent of the chord on the suction surface and 30 percent of the chord on the pressure surface. The ice-protected area extends to only 30 percent of the blade radius. Ice formations are negligible beyond 30 percent of blade radius due to aerodynamic heating and centrifugal force on the outer portions of the propeller blades.

There are three primary penalties of electrothermal deicing systems: weight, electric power, and aerodynamic performance. A secondary concern with this type of system is runback icing.

In some cases, an annunciator light indicates when the system is functioning properly; in other cases, the aircraft's ammeter or power meter is watched for reading changes on system activation.

Note in Fig. 2-7 that the engine inlet area and inboard portion of the wing have electrothermal anti-icing. The reason for electrothermal anti-icing on this portion of the wing is to prevent ice from being ingested by the engines, which would occur with a pneumatic deicing system.

Fluid Ice Protection Systems

Fluid ice protection systems for fixed-wing aircraft were introduced in the mid-1930s. In 1942, TKS Ltd. was formed to meet British government needs. TKS systems went into service toward the end of World War II. Today Aerospace Systems & Technologies, Inc., of Lawrence, Kansas, manufactures the TKS system. The system is certified for many single-engine, light, general aviation airplanes, including Cessna, Beechcraft, Aerospatiale, and Mooney, as well as the Commander 114.

Fluid ice protection systems operate on the principle that the surface to be protected is coated with a fluid that acts as a freezing-point depressant. Current systems normally use a glycol-based fluid. When supercooled water droplets hit the surface, they combine with the fluid to form a mixture with a freezing temperature below the ambient air temperature. The mixture then flows aft under the influence of aerodynamic forces and either evaporates or sheds from the trailing edge of the surface.

Fluid is distributed onto the surface leading edge to be protected by pumping it under pressure through porous material. The use of a fluid can provide anti-icing or deicing protection for almost any surface where fluid can be distributed. The two primary means for accomplishing this are spray nozzles and porous skin panels. The Commander 114 in Fig. 2-8 uses both these applications. The wings and tail surfaces use laser-drilled titanium panels to deliver the fluid. A slinger ring is installed on the propeller, and a spraybar is positioned in front of the windscreen for ice protection, although the fluid from the propeller slinger ring is often enough to protect the windscreen.

When sufficient fluid is present, no ice will form, and the system functions in an anti-icing mode. If icing conditions become too severe, there may be insufficient fluid flow to totally prevent ice formation. When this occurs, ice will begin to form, usually at or near the aerodynamic stagnation point. If fluid continues to be pumped onto the surface, the ice will not be able to bond firmly and will grow until

Fig. 2-8. The use of a fluid can provide anti-icing or deicing protection for almost any surface where fluid can be distributed. (Photo courtesy of Commander Aircraft Company.)

aerodynamic forces are sufficient to sweep if off the wet surface. This process of periodic growth and shedding is referred to as a *natural deicing mode.* The deicing mode is a condition where ice is allowed to build before fluid flow begins, thus allowing ice to accumulate and bond to the surface. With activation of the fluid ice protection system, a flow is introduced between the ice and surface, weakening the bond so that ice will shed by aerodynamic forces.

The spray-nozzle distribution method is used primarily for windscreen deicing. Liquid ice protection of propellers can be achieved by distributing fluid onto the leading edge of the propeller blades. This is accomplished by feeding the fluid into a propeller slinger ring. Centrifugal force allows the fluid to be deposited on the leading edge at the root of each blade. A rubber boot channels the fluid out along the blade leading edge to a point where centrifugal forces are sufficient to keep the blades ice-free.

If icing conditions are anticipated inflight, the system should be activated during the preflight inspection to ensure that fluid is being delivered to the surface of each panel, slinger ring, and windscreen. This also serves to prime the system. Inflight, the system should be activated immediately prior to or on entering icing conditions.

System advantages consist of complete airframe protection, ease of operation, low maintenance, no reduction in performance, and light weight. The principal penalty of the fluid system is the fluid storage requirement, fluid weight, and added fluid checks. Therefore, its use on large airplanes is not typical. In addition, there are environmental considerations related to toxicity and long-term effects of the fluid.

Electroimpulse Systems

Electroimpulse deicing (EIDC) is classified as a mechanical ice protection method. Ice is shattered, debonded, and removed from an aircraft surface by a hammerlike blow delivered electrodynamically. Removal of the ice is aided by turbulence airflow, requiring relatively low electrical energy.

Physically, the system consists of ribbon-wire coils rigidly supported inside the aircraft surface to be deiced but separated from the skin surface by a small air gap. A sudden high-voltage electric current is discharged through the coil. The circuit must have low resistance and inductance to permit the discharge to be very rapid. A strong electromagnetic field forms and collapses, inducing eddy currents in the aircraft skin. The eddy current and coil current fields are mutually repulsive, resulting in a toroidal-shaped pressure pattern on the skin opposite the coil. The force delivered on the skin is so sharp that it produces a sound resembling a metal-on-metal blow. Actual surface deflection is small, but acceleration rapid. The system is not recommended for windscreen deicing or considered applicable to propellers; however, spinners and nose cones can be deiced by this method.

The first nation to use this system was the former Soviet Union, which had a fully equipped aircraft in 1972 and has equipped several transport airplanes since. The system has had extensive NASA testing and limited flight testing in the United States by NASA and Cessna. Other testing has been done by major air carrier manufacturers. BFGoodrich has proposed this type of system for horizontal stabilizer ice protection.

The EIDI system requires low power, is reliable for all types of ice, is weight comparable, requires low maintenance, is runback-free, and is simple to operate. It has had limited use, is not an anti-icing system, and creates considerable noise. Areas of concern are skin fatigue, electromagnetic interference, adverse effects of lightning strikes, and lack of failure analyses.

Lancair plans to offer an electroexpulsive anti-ice protection system to its new Columbia composite. An electromagnetic impulse debonds the ice, and then aerodynamic forces remove it. The system works with various ice depths and layers from frost to clear ice 1 in. thick. The system works on wing and tail surfaces, as well as on engine intakes and other areas requiring ice protection.

Hot-Air Systems

As in all thermal ice protection systems, in hot-air systems, heat is applied to water droplet impact regions. The heat is used to either prevent the freezing of droplets, to evaporate droplets, or to debond the resulting ice formation. Hot-air systems are used on many air carrier and general aviation jet aircraft (Fig. 2-9). Hot-air systems can be either anti-icing or deicing. Four heat sources are available:

- Extraction of engine compressor bleed air
- Compression of ram air using a dedicated compressor
- Recovery of exhaust gas waste heat—piston engines
- Heating of ram air using a fuel-burning combustion heater

Fuel-burning combustion heaters were employed on the Douglas C-124 Globemaster. The airplane appeared to have auxiliary fuel tanks on the wing tips. These were in fact fuel-burning heaters for the airplane's ice protection system.

Fig. 2-9. Hot-air ice protection systems are used to either prevent the freezing of droplets, evaporate droplets, or debond the resulting ice formation.

Hot-air anti-icing systems use heated air to maintain the temperature of the protected surface above freezing throughout an icing encounter. These systems are classified as fully evaporative or running wet. *Evaporative systems* supply sufficient heat to evaporate all water impacting the surface. *Running-wet systems* provide only enough heat to prevent freezing on the surface. Beyond the heated surface, running-wet systems could allow runback ice. For this reason, running-wet systems must be designed carefully to prevent runback in critical locations, e.g., at turboprop or turbojet inlets. For anti-ice systems, heat must remain on throughout the icing encounter. Areas usually protected are engine inlets and wing and empannage leading edges.

In a deicing system, ice accretion is permitted on the surface and removed periodically. The system must apply enough heat to melt the bonding layer of ice. Aerodynamic or centrifugal forces then remove the bulk of the ice.

Operation can be checked on the ground; however, such systems can present a serious overheating problem. Without ambient airflow over the surface, the skin temperature will approach the bleed air temperature, which could damage the surface. Hot-air systems generally are noted for their high reliability and low maintenance.

Advantages of these systems are the easy availability of ice protection energy, ducting used often can be used for other purposes, they are relatively simple, they are easy to maintain, and there is no decrease in aircraft performance during use of the anti-icing system. Cost of installation may be a disadvantage for some systems. Whenever a hot-air system operates in a running-wet condition, there is a danger of freezing water runback.

TKS Ice Protection System for the Commander 114

The following excerpts from the Commander Aircraft Models 114B and 114TC *Supplement to Pilot's Operating Handbook* and/or the *FAA Approved Airplane Flight Manual* are *for reference only,* courtesy of the Commander Aircraft Company.

Section 2
Limitations

1. Flaps are limited to a maximum deflection of 10° when the TKS-equipped aircraft has encountered icing conditions. An icing condition is defined as visually detected ice, or the presence of visible moisture in any form at an indicated temperature of +3°C or below.

2. Minimum ice protection fluid   7.0 U.S. gallons
   for takeoff into icing conditions

3. TKS Ice Protection System   7.0 U.S. gallons
   maximum capacity
   Normal Flow Duration   161 minutes (2.6 gal/hour)
   Maximum Flow Endurance   80.5 minutes (5.2 gal/hour)

Section 2 (*Continued*)

4. The TKS Ice Protection System must be serviced with one of the following fluids:

    TKS 80

    AL5 (DTD 406B)

    TKS R328

5. Only the following solvents are authorized for cleaning the leading edge panels:

    Water (with soap/detergent)

    Approved de-icing fluid (see 2.)

    Aviation gasoline

    Isopropyl alcohol

    Ethyl alcohol

    Industrial methylated spirit

6. Minimum airspeed for   110 KIAS
    operation in icing conditions
    (except for takeoff and
    landing)

7. Checks and inspections specified under Normal Operating Procedures: Preflight Check, Before Starting Engines, and Before Takeoff Check in this supplement must be satisfactorily completed prior to flight into known icing conditions.

Note from Section 2, "Limitations," the flap restriction (10°) and the definition of icing conditions—visible moisture, temperature +3°C or below. The section also specifies the normal and maximum duration of the ice protection system. Minimum airspeed for operations in icing conditions is 110 knots to prevent excess ice forming on the undersides of the airplane. These limitations are a direct reflection of the hazards previously discussed.

Section 3
Emergency Procedures

Flashing red low-pressure lights illuminated (TKS fluid low pressure)

1. Other pump	Select
2. Flow rate	MAXIMUM if required

NOTE

Activate either windshield pump to prime the alternate pump if necessary.

3. Icing condition	Exit as soon as possible
4. TSK Ice Protection System	Monitor operation

Amber high-pressure light illuminated (TKS fluid high pressure)

1. Reset button	Press
2. Icing conditions	Exit as soon as possible
3. TKS Ice Protection System	Monitor operation

Failure of ice protection system or excessive ice accumulation (observed or suspected) on protected airplane surfaces

1. Icing condition	Exit as soon as possible
2. Flaps	Do not extend beyond the approach position
3. Airspeed	Maintain 120 KIAS or greater until final approach and landing
4. Multiply AFM landing distances by a factor of 1.6	
5. Final Approach	Maintain 100 KIAS or greater

Pitot heat annunciator illuminates with pitot heat turned on.

1. Icing conditions	Exit as soon as possible
2. Airspeed	Airspeed may be unreliable with the failure of pitot heat. Rely on pitch attitude, power settings, and rate of climb to safely fly the aircraft

WARNING

Stall speed will increase with the accumulation of ice on the wing and tail leading edges. Expect higher than normal sink rates with power reduction.

Section 3, "Emergency Procedures," also reflects my discussion of icing thus far. The section provides procedures for failing of the ice protection system—exit as soon as possible, limited use of flaps, increased airspeed on final, and increased landing distances.

Section 4
Normal Procedures

Always perform the steps of this section prior to *any* flight where icing may be a possibility. System response at initial startup is a function of the frequency of use. If the system has not been used recently, the time between system activation and complete fluid coverage of the airframe can be several minutes. Always follow the guidelines of step 3 of the Pre-Flight Inspection to assure proper system response when needed.

WARNING

Do not delay activation of the TKS Ice Protection System if icing conditions are encountered. For best operation, the system must be on prior to accumulation of ice on protected surfaces. In order to minimize ice accumulations on unprotected lower surfaces, maintain a minimum speed of 110 KIAS during operations in icing conditions. This will provide an angle of attack that reduces exposure (frontal area) of the lower surfaces to ice accumulation. If one is unable to maintain 110 KIAS at maximum continuous power, a change of altitude and/or course may be necessary to maintain minimum airspeed and/or exit the icing condition.

NOTE

Conditions exist for icing when the indicated outside air temperature is +3°C or below and visible moisture in any form is present.

If icing conditions are encountered, select MAXIMUM flow until the ice is removed, then select NORMAL or MAXIMUM flow as required to prevent ice accumulation.

The TKS Liquid Ice Protection System should not normally be activated in dry, cold air. The ice protection fluid is designed to mix with water impinging on the aircraft surface in normal operation. If dispersed in dry, cold air, the fluid becomes a gel and takes considerable time to clear, particularly on the windshield.

Section 4 (*Continued*)

The windshield pumps cycle for approximately 4 seconds when they are activated. The windshield will take approximately 30 seconds to clear after the spray has ended. Ice should not be allowed to accumulate on the windshield. Activate either of the windshield pumps as necessary to maintain clear forward vision.

Pre-Flight Inspection

1.	Battery switch	ON
2.	TKS Ice Protection System	MAXIMUM (either pump)
3.	Airframe Inspection	
	Wings, tail, propeller, windshield	Verify free of ice
	Fluid tank	Check quantity (full, 7.0 U.S. gallons
		Check cap secure and door closed
	Porous panels	Check condition and security
		Check evidence of fluid from all panels
4.	Propeller	Check evidence of fluid from propeller
5.	Windshield spraybar	Check condition
6.	Wing Ice Light	Check operation
7.	All switches	OFF

Before Starting Engine

1.	TKS Ice Protection System	
	a. Windshield pumps 1 and 2	Check operation
	b. Main pump 1	MAXIMUM, verify steady green light
		NORMAL, verify steady green light
		OFF

Section 4 (Continued)

c. Main pump 2	MAXIMUM, verify steady green light
	NORMAL, verify steady green light
	OFF
d. High-pressure light	Verify not illuminated (if light illuminated, push RESET and verify if it remains extinguished)
2. Pitot/stall warning heat annunciator extinguishes	ON, verify pitot heat
	OFF, verify pitot heat annunicator illuminates

Before Takeoff Check

If icing conditions exist:

1. TKS Ice Protection System	Select pump 1 or 2, NORMAL flow
2. Pitot/stall warning heat	ON
3. Wing ice light	As required

After Takeoff

If icing conditions exist:

1. TKS Ice Protection System	Select pump 1 or 2, NORMAL flow, MAXIMUM if required
2. Windshield De-ice fluid pump	Activate either pump as required
3. Pitot/stall warning heat	ON
4. Heater and defroster	ON
5. Engine alternate air controls	As required
6. Wing ice light	As required
7. Airspeed	Maintain 110 KIAS or greater

Section 4 (Continued)
Cruise Check

If icing conditions exist:

1. TKS Ice Protection System	Select pump 1 or 2, NORMAL flow, MAXIMUM if required
2. Windshield de-ice fluid pump	Activate either pump as required
3. Pitot/stall warning heat	ON
4. Heater and defroster	ON
5. Engine alternate air controls	As required
6. Wing ice light	As required
7. Airspeed	Maintain 110 KIAS or greater

Before Landing Check

If icing conditions exist:

1. TKS Ice Protection System	Select pump 1 or 2, NORMAL flow, MAXIMUM if required
2. Windshield de-ice fluid pump	Activate either pump as required
3. Pitot/stall warning heat	ON
4. Heater and defroster	ON
5. Engine alternate air controls	As required
6. Wing ice light	As required
7. Airspeed	Maintain 110 KIAS or greater until final approach

Final Approach

With residual ice on the airframe:

1. Windshield de-ice fluid pump	Off at least 30 seconds prior to landing

Section 4 (Continued)
After Landing

Check if icing conditions exist:
1. TKS Ice Protection System   OFF
2. Pitot/stall warning heat   OFF
3. Wing ice light   OFF

Stopping Engine

1. Engine alternate air control   Normal

Section 4, "Normal Procedures," again follows my discussion. It outlines the steps that must be taken prior to any flight where icing may be possible.

Section 5
Performance

Airplane performance and stall speeds in clear air are essentially unchanged with installation of the TKS Ice Protection System. Significant climb and cruise performance degradation, range reduction, as well as buffet and stall-speed increase can be expected if ice accumulates on the airframe. Residual ice on the protected areas and ice accumulation on the unprotected areas of the airplane can cause noticeable performance losses, even with the TKS Ice Protection System operating.

Normal Rate of Climb

Installation of the TKS system will result in a 100 fpm loss in rate of climb. Residual ice on unprotected airplane surfaces can cause a loss in rate of climb of an additional 300 fpm. Additional accumulation of ice on the airplane can result in significant loss in normal rate of climb.

Section 5 (Continued)
Balked Landing Climb

Installation of the TKS system will result in a 50 fpm loss in balked landing climb. Residual ice on unprotected airplane surfaces can cause a loss in balked landing climb performance of an additional 150 fpm. Additional accumulation of ice on the airplane can result in a significant loss in balked landing climb performance.

Stall Speeds

Stall speed is not affected by residual ice on unprotected airplane surfaces. Stall speeds increase significantly with even small accumulations of ice on the wing leading edge. The first 1/4 inch of ice accumulation on the wing leading edges causes the most rapid increase in stall speed. Additional ice accumulation on the wing leading edges results in a continued increase in stall speed, although at a less rapid rate.

Expect higher stall warnings with this system for a clean airframe with no ice. The stall vane has been adjusted to provide adequate warning with residual ice around the vane backplate, resulting in stall warning activation as high as 14 KIAS above stall speed for some flap configurations.

Landing

When the aircraft has encountered icing conditions, flap deflection is limited to a maximum of 10°. An icing condition is defined as visually observing ice accumulation or flight in temperatures at or below +3°C when any type of visible moisture is present.

Use the 10° flap landing distance and approach speed data from Table 5-1 for landings with 10° of flap.

Section 5, "Performance," should be no surprise. Pilots must expect loss of performance when ice accumulates on the airframe. Ice on unprotected areas can cause noticeable performance loss, even with the ice protection system operating.

Section 10
Safety Information

FLIGHT IN ICING CONDITIONS

This airplane has been approved for flight in icing conditions as defined in FAR 25, Appendix C, in accordance with the criteria contained in Advisory Circular 23-1419-2, and the applicable requirements of FAR 23. These conditions do not include, nor were tests conducted in all icing conditions that may be encountered (e.g., freezing rain, freezing drizzle, mixed conditions, or conditions defined as severe). Some icing conditions not defined in FAR 25 have the potential for producing hazardous ice accumulations which (1) exceed the capabilities of the aircraft's ice protection equipment, and/or (2) create unacceptable airplane performance. Flight into icing conditions which lie outside the FAR-defined conditions is not specifically prohibited; pilots are advised, however, to be prepared to divert the flight promptly if hazardous ice accumulations occur.

Safe operation in icing conditions is dependent upon pilot knowledge of atmospheric conditions conducive to ice formation, familiarity with the operation, and limitations of the installed ice protection equipment, and the exercise of good judgment when planning a flight into areas where possible icing conditions exist. Flight into areas with known icing conditions should be avoided or limited to the minimum amount of time absolutely necessary. FAR 25 did not envision long duration ice encounters. The intent of the regulation was to allow aircraft to fly through icing conditions. When possible, prolonged operations in icing conditions should be avoided. When icing conditions are encountered, the recommended procedure is to change to an altitude where icing conditions are not present, particularly if it is known that the icing conditions at the present altitude are widespread. Ice accumulations on the airplane increase aerodynamic drag, reduce airplane range, reduce climb performance, and increase stall speed.

Normal operation of the TSK Ice Protection System results in a continuous build and shed of small quantities of ice along the leading edges. Engine alternate air should be selected as required. The primary indication of loss of primary induction air will be a drop in indicated manifold pressure.

Section 10 (Continued)

To achieve the best visibility, a straight-in approach should be utilized whenever possible if ice has accumulated on the right windshield and unprotected areas of the left windshield. The windshield pump should be off at least 30 seconds prior to landing to allow adequate time for the windshield to clear.

Accumulation of ice on unprotected lower surfaces is minimized by maintaining a minimum airspeed of 110 KIAS, until a lower speed is required for final approach and landing. This speed provides an angle of attack that minimizes exposure (frontal area) of lower airframe surfaces to ice accumulation. The pilot should take appropriate actions to maintain this minimum speed, including increasing power (up to maximum available if necessary), change of altitude, descent, etc. Prolonged operation at lower speeds may result in substantially greater performance losses than specified in Section 5 of this supplement.

By definition, icing conditions are considered to exist when the indicated outside air temperature is below +3°C and any kind of visible moisture is present. Outside air temperature should be closely monitored when flying in clouds or precipitation. The most significant icing, found in stratus type clouds, is generally located near the top of a well-defined cloud formation. Severe icing conditions exceeding the capability of the ice protection system can be encountered in many different situations. Examples of these conditions include rapidly building cumulus clouds, up-slope environments, etc.

The prudent pilot must remain alert to the possibility that icing conditions may become so severe that the TKS ice protection equipment cannot cope with the situation. If such a condition is encountered, the pilot should immediately take the most safe and expeditious course of action to exit the condition.

Section 10, "Safety Information," is a summary of the limitations and procedures from previous sections. It cannot be overemphasized that approved ice protection systems are not designed or tested for all icing conditions.

ICING MYTH

All we really need is prop deice.

CHAPTER 3

Icing Reports and Observations

Many have touted the need for pilots to get the "big picture." This refers to obtaining a complete weather synopsis, i.e., the position and movement of weather-producing systems and those which pose a hazard to flight operations. This is important, but it is only one of the elements needed for an informed weather decision. I prefer the term "complete picture." As well as a thorough knowledge of the weather synopsis, pilots must be able to obtain and understand all the information available from current weather reports—discussed in this chapter—and forecast weather—presented in Chapter 4. As we shall see, each report, chart, or product provides a clue to the "complete picture." However, each must be interpreted with a knowledge of its own scope and limitations. Then, with a knowledge of the "complete picture," a pilot can apply the information to a specific flight.

Icing reports and observations consist of surface observations, observations from pilots—either on the ground or during flight—various weather charts, and satellite images. The three-dimensional observational system begins with surface observations—the lower layer; next comes upper-air observation, radar, and PIREPs—the middle layer; and finally, satellite images provide a look from the top down. Additionally, "observations" of airport surface conditions are reported and disseminated to pilots in the form Notices to Airmen (NOTAMs).

Because of rivalry and failure of the Federal Aviation Administration (FAA) and the National Weather Service (NWS) to take prompt action, it has taken over 2 years to adopt the international standard to the PIREP format. The new standard became effective on August 13, 1998. With its adoption, all reports and forecasts used within the United States have been standardized.

> **FACT**
>
> METAR (Meteorological Aviation Routine), SPECIs (Special Reports), and TAFs (Terminal Aerodrome Forecasts) have been used in the United States since July 1, 1996. Mexico converted in 1995 and Canada in 1996. Ostensibly, METAR has standardized surface aviation reports and forecasts worldwide.

Weather charts contain reported or forecast conditions, often computer generated and analyzed. Observed data are derived from METAR, upper air observations, NEXRAD weather radar, and satellite images. In the charts portion of this chapter we will discuss products that contain observed data.

Satellite images report conditions as viewed from space. They tell us about cloud tops and types of clouds and often fill in the gaps between surface-weather reporting locations.

The final section of this chapter deals with Notices to Airmen (NOTAMs). Personnel responsible for airports report surface conditions through the FAA's NOTAM system. Like METAR, TAF, and PIREPs, NOTAMs are now formatted in the international standard. Like weather products, NOTAMs must be decoded, translated, and interpreted to be understood and used effectively.

METAR and SPECI

With METAR and TAF, individual member countries are allowed to change certain items within their reports. The United States continues to report wind in knots, visibility in statute miles, cloud heights in feet, and the altimeter setting in inches of mercury—rather than their metric equivalent. Temperatures, however, are reported in degrees Celsius. In METAR and TAF, the letter *M* means minus or less than, and *P* means plus or more than.

METAR reports appear in the following sequence: Missing or not reported elements are omitted.

Type of Report

There are two types of reports: METAR, a routine observation, and SPECI, a special weather report. METAR observations are reported each hour. Normally, elements of the report are observed between 45 and 55 minutes past the hour. Whenever a significant change occurs, a SPECI is generated. Complex criteria determine the requirement for SPECIs. Generally, they are required when the weather improves to or deteriorates below Visible Flight Rules (VFR) or approach and landing Instrument Flight Rules (IFR) minimums. SPECIs are also required for the beginning, ending, or change in intensity of freezing precipitation or thunderstorm activity.

- Type of report
- Station identifier
- Date and time of report
- Report modifier, if required
- Wind
- Visibility
- Weather and obstructions to visibility
- Sky condition
- Temperature and dew point
- Altimeter setting
- Remarks

SPECIs are not available for all locations; however, these reports normally will carry the remark NOSPECI. In such cases, significant changes can occur without expeditious notification.

Station Identifier

METAR uses standard four-letter International Civil Aviation Organization (ICAO) location identifiers (LOCID). For the continental United States, this consists of three letters prefixed with the letter *K*. For example, Newark, New Jersey, is *KEWR* and Philadelphia, Pennsylvania, is *KPHL*. (Prior to U.S. acceptance of the international standard, LOCIDs for weather reports were alphanumeric. That is, made up of letters and numbers (O45—Vacaville, California). Now, with more and more weather reports transmitted over FAA and NWS communication systems, all locations eventually will use letters only (*KVCB*—Vacaville, California). This is why we have seen so many changes to LOCIDs recently and why they are making even less sense than they used to.

Official LOCIDs are contained in FAA Handbook 7350.5, *Location Identifiers,* available for sale from the Superintendent of Documents. Pilots also have access to LOCIDs from aeronautical charts and through FAA Flight Service Stations (FSSs), Direct User Access Terminals (DUATs), and the *Airport/Facility Directory*—the green book.

Date/Time Group

The time of observation is transmitted as a six-digit date/time group appended with a *Z* to denote Coordinated Universal Time (UTC), at times—pardon the pun—referred to as *ZULU* or *Z.* [It seems that an advisory committee of the International Telecommunications Union in 1970 was tasked with replacing the international time standard of Greenwich Mean Time (GMT). The question became whether to use English or French word order for Coordinated Universal Time—sound familiar? So instead of CUT or TUC, UTC was adopted and became effective in the late 1980s.] The first two digits represent the day of the month, and the last four digits represent the time of observation.

Report Modifier

Two report modifiers may appear after the date/time group. *AUTO* indicates that the report comes from an automated weather observation station that is not augmented. *Augmented* means someone is physically at the site monitoring the equipment. It does not, however, mean that a real person is changing, correcting, or adding information to the observation. Automated observation, identified by *AUTO* in the report modifier or *AO*...in remarks, means sky condition and visibility are not observed in the same manner as manual (human) observations.

COR means the METAR/SPECI report originally was transmitted with an error that now has been corrected. The only way to identify the correction is to compare the report with the original.

Wind

Wind is reported as a 2-minute prevailing wind direction and speed using a five-digit group, six if the speed exceeds 99 knots. The first three digits indicate the direction from which the wind is blowing, in

10° increments, in relation to true north. [The only time pilots can expect to receive wind direction in relation to magnetic north is from a control tower, an FSS providing a Local Airport Advisory (LAA), an Automatic Terminal Information Service (ATIS) recording, or an ASOS/AWOS radio broadcast.] The next two or three digits indicate speed. The units of measurement follow: KT = knots; KMH = kilometers per hour; and MPS=meters per second. As mentioned previously, in the United States, knots will continue to be used (*34017KT* means wind at 340° from true north at 17 knots). Gustiness is reported with a *G* (*23020G30KT*). Gusts refer to rapid fluctuations in speed that vary by 10 knots or more. If wind direction varies by 60° or more with a speed greater than 6 knots, a variable group separated by a *V* follows the prevailing group (*34017KT 310V010*). A calm wind is reported as *00000KT.* The contraction *VRB* indicates a variable wind direction. *VRB* may be included for winds less than 7 knots (*VRB04KT*). *VRB* also may be used in special cases at higher speeds, such as a wind shift with the passage of a thunderstorm over the station.

Since runways are identified by their magnetic orientation, wind direction and speed are crucial for determining crosswind component. This is especially true in areas with large magnetic variation. This may be critical with slush-, snow-, or ice-covered runways, taxiways, and ramps. In regions of freshly fallen snow, strong winds (20 to 30 knots) can lift the snow, often reducing surface visibility to zero.

Visibility

In the United States, visibility from manual observations or its automated station equivalent is reported as prevailing in statute miles, designated *SM*. For example, *1/2SM* means one-half statute mile, *7SM* means seven statute miles, or *15SM* means fifteen statute miles. For automated stations, a visibility of less than one-quarter mile is reported *M1/4SM,* and maximum reported visibility is *10SM.*

One NWS observer was quite perplexed when the tower always reported increased visibility after sunset. The reason

VISIBILITY

Visibility is a measure of the transparency of the atmosphere. During the day, manually reported visibility represents the distance at which predominant objects can be seen; at night, visibility is the distance that unfocused lights of moderate intensity are visible.

was the change in criteria for the observation. Pilots should note that daytime values do not necessarily represent the distance that other aircraft can be seen. At night, especially under an overcast, unlighted objects may not be seen at all, and there could be no natural horizon. Variable visibility has the same implications as variable sky condition and ceiling—conditions changing rapidly at the airport.

Runway visible range (RVR), when reported, uses the following format. The letter *R* followed by the runway number, a solidus /, and the RVR in feet, e.g., *R29/2400FT* (meaning runway 29, visible range 2400 feet). The following will be added to the visible range as required:

- *V*, variability (*R32R/1600V2400FT*)

- *M*, less than (*R22R/M1600FT*)

- *P*, more than (*R36/P6000FT*)

Reported surface visibility comes into play when a pilot plans to take off or land or enter the traffic pattern under the provisions of VFR within Class B, C, D, and E airspace. Surface visibility must be at least 3 miles. If surface visibility is not reported, flight visibility must be at least 3 miles.

METAR and SPECIs report surface visibility, which does not necessarily represent conditions at altitude. Visibilities aloft most often are reduced by rain, snow, dust, smoke, and haze. Pilots flying either VFR or IFR must be prepared for reduced visibilities in freezing or frozen precipitation. Pilot reports are the only source of flight visibility information.

Weather and Obstructions to Vision

In METAR, weather and obstructions to vision may contain some or all of the following: intensity, proximity, descriptor, precipitation, obstruction to vision, and other. Weather and obstructions to vision and their codes follow.

Table 3-1 Weather and Obstructions to Vision

Intensity	
—	Light
no symbol	Moderate
+	Heavy
Proximity	
VC	Vicinity
Descriptor	
TS	Thunderstorm
DR	Low Drifting
SH	Showers
MI	Shallow-*mince*
FZ	Freezing
BC	Patches-*banc*
BL	Blowing
PR	Partial
Precipitation	
RA	Rain
DZ	Drizzle
GR	Hail ($\geq$ than $\frac{1}{4}$ in.)-*grêle*
GS	Small Hail/Snow Pellets-*grésil*
SN	Snow
PE	Ice Pellets
SG	Snow Grains
IC	Ice Crystals
UP	Precipitation (Automated Observation)
Obstructions to Vision	
FG	Fog (Visibility less than $\frac{5}{8}$)
PY	Spray
BR	Mist (Visibility $\frac{5}{8}$ to 6)-*brume*
SA	Sand
FU	Smoke-*fumée*
DU	Dust
HZ	Haze
VA	Volcanic Ash

Table 3-1 Weather and Obstructions to Vision (*continued*)

Other

SQ	Squall
SS	Sandstorm
DS	Duststorm
PO	Dust/Sand Whirls
FC	Funnel Cloud
+FC	Tornado
+FC	Waterspout

It is important to remember that intensity refers to the precipitation, not the descriptor. For example, +TSRAGR is a thunderstorm with heavy rain and hail. A severe thunderstorm is indicated by surface winds of 50 knots or hail $3/4$ in. in diameter.

Proximity or vicinity, designated *VC*, reports weather occurring in the vicinity of the airport. Vicinity is defined for precipitation as not occurring at the point of observation but within 10 statute miles of the station; when used to report any type of fog, vicinity means from between 5 and 10 statute miles.

Descriptors apply to precipitation or obstructions to vision. Automated sites may use *UP* to report precipitation of unknown type. Fog (*FG*) is only reported when the visibility is less than $5/8$ mile. With visibility of $5/8$ mile or greater, mist (*BR*) is used. In METAR, mist (*BR*) refers to an obstruction to vision, not precipitation.

Five categories of weather phenomena appear in the "other" group. A squall (*SQ*) is a strong wind characterized by a sudden onset, a duration of minutes, and then a sudden decrease in speed. Sandstorms (*SS*) and duststorms (*DS*) are differentiated from blowing dust and blowing sand by visibility. Sandstorms and duststorms only appear when the visibility is less than $5/8$ mile. Dust or sand whirls (*PO*) are dust or sand raised by rapidly rotating columns of air. When well developed, they are *dust devils.*

Recall from previous chapters that freezing rain (*FZRA*), freezing drizzle (*FZDZ*), and ice pellets (*PL*) alert pilots to significant icing conditions, snow pellets (*GS*) to a lesser degree. As well as an icing hazard, hail (*GR*) implies possible structural damage to the aircraft.

Snow (*SN*) typically does not produce a serious icing hazard, but accumulations on airport surfaces present an altogether different hazard. Blowing snow (*BLSN*) reduces visibility, creating a different type of hazard, especially for VFR pilots. Blowing snow can reduce visibility to near zero but typically is restricted to within a few hundred feet of the surface. Visibility improves rapidly when the wind subsides. Drifting snow (*DRSN*) is raised by the wind, remains close to the surface, and does not significantly reduce visibility. Snow drifts, which can be inferred by reports of drifting snow, present an airport surface condition hazard that typically is advertised in a NOTAM.

Sky Condition

METAR sky condition reports consist of the amount of cloud coverage, height above ground level (AGL) of the layer, and, under certain conditions, cloud type: towering cumulus (*TCU*) and cumulonimbus (*CB*).

The amount of sky cover is reported in eighths—sometimes referred to as *octas*—using the following contractions:

- *CLR*, clear below 12,000 feet (automated reports)

- *SKC*, clear (no clouds)

- *FEW*, few ($>1/8$ to $2/8$ coverage)

- *SCT*, scattered ($3/8$ to $4/8$ coverage)

- *BKN*, broken ($5/8$ to $7/8$ coverage)

- *OVC*, overcast ($8/8$ coverage)

AT A GLANCE	
BKN	broken clouds
BLSN	blowing snow
BR	mist
CB	cumulonimbus
CLR	clear below 12,000 ft.
DRSN	drifting snow
DS	duststorm
FEW	few clouds
FG	fog
FZRA	freezing rain
GR	hail
GS	snow pellets
OVC	overcast
PL	ice pellets
PO	dust or sand whirls
SCT	scattered clouds
SKC	clear; no clouds
SN	snow
SQ	squall
SS	sandstorm
TCU	towering cumulus
UP	unknown precipitation type

When a cloud layer develops below the point of observation, such as Mt. Wilson above the Los Angeles Basin, the layer is encoded *OVC///* (an overcast layer with tops below the point of observation).

A *partial obscuration* is reported when between one-eighth and seven-eighths of the sky is hidden by a surface-based obscuring phenomenon. Precipitation—including snow, haze, smoke, and fog—usually causes this condition.

In Fig. 3-1, half, or four-eighths, of the sky is hidden by fog. The observer sees another one-eighth cloud cover at 3000 ft. In the METAR code, a partial obscuration is indicated as *FEW, SCT,* or *BKN* on the surface (*FEW000,* between one-eighth and two-eighths of the sky obscured). In Fig. 3-1 our observer reports visibility reduced by fog (*FG*), between three-eighths and four-eighths of the sky obscured, with a ceiling of 3000 ft broken. (The observer used the summation principle of cloud cover; four-eighths plus one-eighth equals five-eighths. Five-eighths is reported as a broken layer. The remark tells us that it is an obscuration (*...RMK FG SCT000*). What else could it be? Well, technically, a layer with a base of less than 50 ft would be reported as *000*. This is very unlikely. If the observer did indeed mean to report a layer with a base of less than 50 ft, the remark *FG SCT000* would not appear.

Why such a complicated procedure to report a partial obscuration? The international METAR code has no provision for reporting this phenomenon. The FAA and NWS concurred—probably for the first time—and proposed that a partial obscuration be eliminated from U.S. reporting procedures. However, there are other players in the game, namely, the Department of Defense (DOD). Within DOD is the Department of the Navy and the Marine Corps. Well, the Marines just could not get along without a partial obscuration. Semper fi! (As of this writing, the FAA and NWS are still attempting to eliminate partial obscurations from reporting criteria.)

Normally, a pilot can expect ground contact while flying in areas with a reported partial obscuration. This is why they are not

PARTIAL OBSCURATION

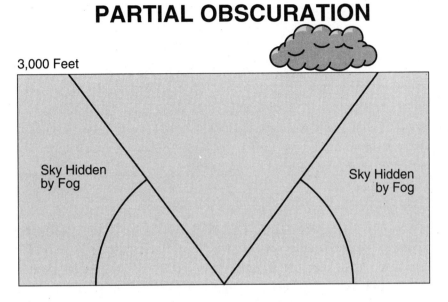

METAR: FG SCT000 BKN030
...RMK FG SCT000

Fig. 3-1. A partial obscuration is reported when between one-eighth and seven-eighths of the sky is hidden by a surface-based obscuring phenomenon.

considered ceilings. Penetrating a partial obscuration VFR requires the appropriate horizontal visibility for the class of airspace.

In METAR, ceiling is not designated. For aviation purposes, the *ceiling* is the lowest broken or overcast layer or vertical visibility into a total obscuration; a partial obscuration does not constitute a ceiling.

Cloud heights are reported with three digits, in hundreds of feet above ground level (AGL). When towering cumulus (*TCU*, cumulus congestus) or cumulonimbus (*CB*) clouds are present, they will be included in the sky condition section of the report (*BKN035TCU* means towering cumulus clouds with a ceiling at 3500 ft broken).

Vertical visibility into a total obscuration will be reported using the letters *VV* followed by the vertical visibility in hundreds of feet AGL

(*VV002* means indefinite ceiling at 200 ft). VFR pilots must use caution when flying in a total obscuration. For example, consider the following situation: *...3SM -SN VV015....*The visibility is 3 mi in light snow, indefinite ceiling at 1500 ft. Is it VFR? Technically, yes. A pilot could expect ground contact at 1500 ft, but slant-range visibility will be considerably less. Additionally, should the pilot climb above 1500 ft AGL, he or she may not be in clouds but will have no visible contact with the ground and no natural horizon! This is a good example of a situation that may be legal but not safe.

Temperature, Dew Point, and Altimeter

Temperature and dew point are reported using two digits in degrees Celsius. Temperature below zero will be prefixed with the letter *M*. For example, *20/15* means the temperature is 20°C and the dew point is 15°C, and *08/M03* means the temperature 8°C and the dew point is −3°C.

Reports of temperature are important beyond personal comfort value. High and low temperatures affect aircraft operation and performance. Temperatures at or below freezing imply that any precipitation will be of the freezing or frozen variety, with all the accompanying icing implications. An approach and landing under these conditions mean that any ice accumulation will be carried to the ground, which may be covered by slush, snow, or ice. Expect morning frost with a forecast of clear skies, close temperature–dew point spread, light surface winds, and below-freezing temperatures.

The altimeter, in inches of mercury, follows temperature and dew point in a four-digit group prefixed with the letter *A*. For example, *A2992* means altimeter setting at 29.92 inches of mercury.

Remarks

Remarks follow the contraction *RMK*. Remarks are divided into automated, manual, and plain language and additive and automated maintenance data. Automated remarks indicate the type of automated station. Manual remarks supplement information in the body of the report. Plain-language remarks contain other significant data. For example, FIRST or LAST observation of the day. Additive data are used by the NWS and consist of climatologic information, usually in numerical code groups. Automated maintenance data consist of sensor outages and maintenance requirements. Remarks may include significant types of precipitation, such as WET SNOW.

The following are examples of METAR reports:

SPEC KEWR 172240Z AUTO 03545G60KT 0SM R04/4500FT +TSRA SQ VV000 23/21 A2989 RMK A02 TS OVHD MOVG ESE FQT LTGICCCCG

Decode: Special report for Newark, New Jersey, taken on the 17th day of the month at 2240 UTC by a nonaugmented automated weather observation station; wind 030° at 45 knots with gust to 60 knots; visibility zero, runway 4 visible range 4500 ft in thunderstorm with heavy rain and squalls; indefinite ceiling zero; temperature 23°C, dew point 21°C; altimeter 29.89; remarks: automated station with precipitation sensor, thunderstorm overhead moving east-southeast, frequent lightning in cloud, cloud to cloud, and cloud to ground.

METAR   KPHL   172250ZCOR   36018G24KT   1SM R27R/3000VP6000FT +TSRA VV006 22/20 A2983 RMK TSB25 ALQDS MOVG SE FQT LTGICCG RAB38 PRESFR

Decode: Routine hourly observation for Philadelphia, Pennsylvania, taken on the 17th day of the month at 2250 UTC, the report has been corrected; wind 360° at 18 knots with gusts to 24 knots; visibility 1 statute mile, runway 27 right visible range variable between 3000 and more than 6000 ft; thunderstorm with heavy rain; indefinite ceiling at 600 ft; temperature 22°C, dew point 20°C; altimeter setting 29.83; remarks: thunderstorm began at 25 minutes past the present hour, thunderstorm all quadrants moving southeast with frequent lightning in cloud and cloud to ground, rain began 38 minutes past the present hour, the pressure is falling rapidly.

PIREPs

Ever complain about forecasts, think they were prepared in a sterile environment, or conclude that pilots have no influence on their preparation? If pilots wish to participate, they can and do influence forecasts. According to the National Weather Service (NWS), PIREPs are the most important ingredient for the AIRMET Bulletin, SIGMETs, Center Weather Advisories (CWAs), and Winds and Temperatures Aloft forecast amendments. In addition to the influence on forecasts, certain phenomena such as cloud layers and tops, icing, and turbulence can only be observed and reported by pilots.

The FAA, recognizing the importance of PIREPs, has directed air traffic controllers to actively solicit PIREPs, especially during marginal or IFR conditions and periods of hazardous weather. Pilots operating IFR must by regulation "...report...any unforecast weather conditions encountered...." These reports are of value not only to other pilots but also to center, terminal, and flight service controllers and forecasters alike.

PIREPs can be provided to any air traffic facility (center, tower, FSS). However, to ensure widest distribution, it is best to report directly to flight service, preferably Flight Watch—En-route Flight Advisory Service (EFAS).

PIREPs are transmitted under the location identifier for the surface report (METAR) nearest the occurrence using the file type *UA* (*SAC UA* means Sacramento pilot report). Reports may be appended, however, to major hub locations to ensure greatest prominence and widest distribution.

OK, that's the commercial. Now we can get down to the specifics of this often underutilized, most valuable product.

A popular notion in some aviation circles is that a pilot's mere mention of ice will receive emergencylike handling. Icing may be an emergency, but remember, the controller's job is to separate aircraft within a finite amount of airspace. Air traffic control (ATC) may have to assign a higher altitude, but ATC cannot, and should not, be expected to fly the aircraft or assume the responsibility of the pilot-in-command. To paraphrase: An accurate report of actual icing conditions is worth a thousand forecasts.

PIREP Format

Because of lack of direction among both FAA and NWS personnel, PIREPs have appeared in confusing, misleading, and nonstandard formats. This has led to considerable misunderstanding and, in fact, has had an impact on the safety of our ATC system. Correctly formatted PIREPs will eliminate any confusion and increase the usefulness of this valuable product.

PIREPs are entered using the standard format illustrated in Fig. 3-2. There is no need to memorize the form because FAA/NWS specialists encode the report. However, an understanding of the format illustrates the information needed and will assist in decoding and interpreting reports. Standard contractions are used.

Report Type: **UUA or UA**   An urgent pilot report (*UUA*) represents a hazard or potential hazard to flight operations. The message type UUA receives special handling and immediate distribution. Urgent PIREPs contain information on tornadoes, severe or extreme turbulence, severe icing, hail, low-level wind shear when the pilot reports an airspeed change of 10 knots or more, volcanic ash, or any other phenomena considered hazardous. Others, designated *UA*, receive routine distribution.

Location: **/OV**   This is the location where the phenomenon was observed in reference to a VHF NAVAID or using a three- or four-letter identifier. References to other geographic locations may appear in remarks.

Time: **/TM**   This is the time (UTC) the phenomenon was observed.

Altitude/Flight Level: **/FL**   This is the altitude, in hundreds of feet above mean sea level (MSL), at which the phenomenon was encountered.

PIREP FORM

Pilot Weather Report   ➤ = Space Symbol

3-Letter SA Identifier

1. **UA** ➤ ____ **UUA** ➤ ____

___ ___ ___ ➤   *Routine*   *Urgent*
                                 Report   *Report*

	Location:
2. **/OV** ➤	
3. **/TM** ➤	Time:
4. **/FL**	Altitude/Flight Level:
5. **/TP** ➤	Aircraft Type:

Items 1 through 5 are mandatory for all PIREPs

	Sky Cover:
6. **/SK** ➤	
7. **/WX** ➤	Flight Visibility and Weather:
8. **/TA** ➤	Temperature *(Celsius):*
9. **/WV** ➤	Wind:
10. **/TB** ➤	Turbulence:
11. **/IC** ➤	Icing:
12. **/RM** ➤	Remarks:

FAA FORM 7110-2 (1-85) Superscedes Previous Edition

Fig. 3-2. The FAA's PIREP form illustrates needed information and will assist in decoding and interpreting pilot weather reports.

DURGC (during climb) or *DURGD* (during decent) may appear in remarks.

Type Aircraft: /TP This is the type of aircraft using standard international (ICAO) aircraft type designators. From time to time, this element may contain */TP PUP* (pickup truck), */TP CAR*, */TP FBO* (airport fixed base operator), */TP MAN*, or perhaps */TP DON* (Donald at the Half Moon Bay Airport).

Sky Cover: /SK Standard sky cover contractions (*SKC, FEW, SCT, BKN, OVC*) are used followed by cloud bases in hundreds of feet—like METAR, TAF, and aviation forecasts. It is important to remember that PIREP bases and tops are always reported in reference to mean sea level (MSL). Cloud tops are indicated by the word *TOP* followed by the height—similar to aviation forecasts. Additional layers are separated by a solidus (/), and clear above is indicated by *SKC*.

For example:

/SK FEW-SCT-TOP030/BKN060-TOP100/SKC

Decoded: A few to scattered ($^1/_8$ to $^4/_8$ coverage) clouds with tops at 3000 ft MSL; a broken layer ($^5/_8$ to $^7/_8$ coverage) bases at 6000 ft MSL, tops at 10,000 ft MSL; clear above.

Weather: /WX Weather encountered and flight visibility are reported in this element. Flight visibility is reported to the nearest whole statute mile and encoded with the suffix *SM*. Unrestricted visibility appears as *FV99SM*. Weather, when reported, follows visibility using standard ICAO contractions shown on pp. 105 and 106. Should hail be reported, size appears in remarks in $^1/_4$-in increments (*GR1/2* means hail that is $^1/_2$ in thick).

Air Temperature: /TA Air temperature is reported in degrees Celsius, with negative values prefixed with the letter *M*—like METAR.

Wind: /WV Wind direction and speed are encoded using three digits to indicate direction and two or three digits to indicate speed in knots, followed by the letters *KT*—again like METAR and TAF.

Turbulence and icing are transitory and difficult to forecast. Although these terms are defined in the *Aeronautical Information Manual* and various other publications, reports of turbulence and icing are subjective. Despite some accounts to the contrary, there is a tendency for pilots, especially new or low-time pilots, to overestimate intensities of turbulence and icing. Unfortunately, this practice tends to skew forecasts, since forecasters must take them at face value. Pilots who believe they have safely negotiated severe conditions may ignore accurate warnings—another negative effect of overestimating intensities. PIREPs that are not objective are worse than useless.

Turbulence: **/TB** The intensity (*NEG, LGT, MOD, SEV, EXTRM*) of turbulence appears in this element. (Altitude is included when different from the */FL* field.) *CAT* (clear air turbulence) or *CHOP* are entered when reported. Should turbulence be forecast, but reports indicate smooth, */TB NEG* appears. */TB NEG* is interpreted as smooth. *NEG* reports are as important as those reporting turbulence.

Icing: **/IC** The intensity (*NEG, TRACE, LGT, MOD, SEV*), type (*CLR, RIME, MX*), and altitude when different from */FL* are entered in this element. Like turbulence, when icing is forecast but not encountered, *NEG* is entered. Temperature also should be included with icing reports. Like turbulence, a *NEG* encounter is as important and useful as one reporting the phenomenon.

Remarks: **/RM** This element reports low-level wind shear, convective activity, dust and sandstorms, surface conditions at airports, or other information to expand or clarify the report. To be of most value, reports must accurately contain location, time, altitude, type aircraft, sky cover, and temperature, as well as turbulence and icing. Pilots should make every effort to provide complete and timely information.

At present, most domestic PIREPs are entered manually by FSS controllers, Center Weather Service Unit (CWSU) meteorologists, Air Route Traffic Control Center (ARTCC) weather coordinators, and military base operations personnel. To clarify any miscoded reports, pilots have only one option: Contact the local FSS. However, if the local FSS did not input the report, there may be no way to verify the information. In such cases, the only option is to ignore the report.

Pilots must evaluate PIREPs within the context of all available reports and forecasts (part of the "complete picture"). For example, a single report of severe turbulence under clear skies and light winds at the surface and aloft should be viewed with skepticism. On the other hand, a report of severe turbulence with conditions favorable for mountain wave activity, with advisories in effect, must be taken seriously.

Should you hear someone complain about the lack of weather information or forecast accuracy, ask if he or she routinely provides timely, objective pilot reports. Reports confirming the forecast are as important as those for unforecast weather. To obtain the maximum benefit from the aviation weather system, all pilots must get involved. The next time you fly and just do not get around to providing a PIREP, consider the following:

BFL UA /OV BFL /TM 1450 /FLUNKN /TP ALL /RM WISH I HAD A TOP REPORT FROM BFL TO ONT!

The following are two examples of pilot weather reports:

UA /OV OAK-OAK 060080/TM 1345/FL080/TP P28T /SK FEW-SCT030-TOP045/OVC060-TOP070/SKC/WX FV05SM HZ /TA M03/WV24513KT/TB OCNL LGT/MOD 030-070 /IC TRACE-LGT RIME 060-070/RM BTN OAK AND COLUMBIA (022)

This is a routine report from Oakland, California, to 80 nautical miles on the Oakland VORTAC 060 radial. Time of the report is 1345Z. Cruising altitude 8000 ft MSL in a Piper Turbo-Arrow. There are less than $1/8$ to $3/8$ to $3/8$ to $4/8$ coverage, bases at 3000 ft MSL, tops at 4500 ft MSL; $8/8$ coverage, bases at 6000 ft MSL, tops at 7000 ft MSL; clear above. Flight visibility is 5 statute miles in haze, temperature $-03°C$, wind from 345° at 13 knots. The pilot reports occasional light to moderate turbulence between 3000 and 7000 ft MSL; a trace to light rime icing between 6000 and 7000 ft MSL. Remarks report the flight was from Oakland to Columbia.

ASE UUA /OV ASE 090060-ASE 270010/TM 1235/FL140/TP MU2 /SK OVC UNKN-TOP190/IC SEV 180-140/RM 1/4 IN EVERY 30 SEC UNABLE TO MAINTAIN ALT UNTIL 140.

Talk about an exciting ride! This MU2 pilot between 60 east of Aspen, Colorado, and 10 west reports tops at 19,000 ft, icing severe between

18,000 and 14,000 ft. One-quarter inch of ice accumulated every 30 seconds. The pilot was unable to maintain altitude until descending to 14,000 ft.

CASE STUDY

A nonturbocharged, non-deiced Baron departed Reno, Nevada, for southern California. Moderate icing and severe turbulence were forecast. The pilot elected to fly a direct course along the crest of the Sierra Nevada Mountains, the route where the most intense icing and turbulence could be expected. The aircraft iced up with fatal results.

The pilot had no way out because the MEA was the airplane's service ceiling. The terrain was well above the freezing level. The pilot failed to reverse course at the first sign of ice. What other options were available? The pilot could have crossed the mountains near Sacramento, minimizing exposure to ice, but once over the Sierras, it was all downhill. The pilot could have flown toward Las Vegas, where the weather was considerably better, or simply waited for better weather.

When the aircraft became covered with ice, the pilot had no option but to ride it to the crash site.

CASE STUDY

A Bonanza pilot departed the San Francisco Bay area on a flight to Los Angeles. Icing above 7000 ft was forecast and reported. The pilot elected to fly at 11,000 ft. The pilot's last words were, "I've iced up and stalled." The crash occurred in the San Joaquin Valley, where the elevation was near sea level. Minimum altitudes in the vicinity of the crash were well below the freezing level. The pilot simply did nothing until airplane control was lost. In both instances, an ominous PIREP describing icing might read: /IC FATAL.

No one has any business flying in these conditions:

FAT UUA /OV FAT 090030…/FL160-240 /TP F18/IC SEV CLR BFL UUA /OV PMD 330040…/FL100 /TP C402/IC SEV RIME/RM PUP 1 INCH CANT SEE THRU WINDSHIELD

PIREPs from air carriers, the military, and corporate aircraft tend to be more accurate because of the pilots' training and experience. Few student pilots fly DC-10s or F-14s. This is not to say PIREPs from pilots of Cessna 150s or Piper Tomahawks are never accurate and should be ignored.

Pilots must evaluate PIREPs within the context of surface reports, forecasts, and other PIREPs. A single report of severe icing from a Beech Sundowner under conditions not conducive to icing should be viewed with skepticism. On the other hand, a report of severe icing from a Cessna 421 with conditions favorable for icing and advisories in effect must be taken very seriously.

Every time we fly, we become observers, but reports must be timely. Some pilots have a tendency to wait until the latter portion or end of a flight to provide a report. A pilot on a flight from Seattle to Los Angeles contacted Oakland Flight Watch and reported conditions departing Seattle 2^1/$_2$ hours earlier. A somewhat overzealous FSS briefer instructed a student to make a pilot report at the conclusion of the flight. The student, calling the FSS the following day, meekly apologized for failing to provide the report and then proceeded to recount in detail conditions encountered. I would bet that this pilot does not forget on the next flight. Get into the habit of routinely providing timely PIREPs; reports confirming the forecast are as important as those for unforecast weather.

FSS and NWS specialists sometimes tend to editorialize on PIREPs, usually around the time of championship sporting events. Although such reports are unauthorized and unprofessional, pilots can expect to see these, usually humorous, reports from time to time.

Other comments contain personal, social, or political messages, such as the following example:

```
HWD US /OV OAK 110007/TM1600/FL060/TP BE33/SK SKC/WX
FV99SM/TB NEG/RM DURGC HWD NBND SVRL H LYRS AT
FL015/FL042/FL060. SLANT VSBY 15-30MI. PTCHY ST OFSHR-THRU
```

GOLDEN GATE OVR CITY OF OAKLAND. REPORTED BY A DERANGED BONANZA PILOT

OK, I was the deranged Bonanza pilot!

Charts

The general and even some specific locations of potential icing can be determined by analyzing various current weather charts. This section discusses surface analysis and weather depiction charts, radar summary charts, upper-level or constant-pressure charts, and the charts that make up the composite moisture stability series—lifted-index analysis, precipitable water, freezing level, and average relative humidity. For our purposes, the discussion of charts will be restricted to their icing implications.

A WORD OF CAUTION

Like current weather reports, observed data charts always show the state of the atmosphere in the past—where the weather's been. They can never be used alone or in lieu of forecast products. They are, however, a good starting point.

The *surface analysis chart* provides a first look at weather systems. The *weather depiction chart* analyzes surface weather into three major weather categories: IFR, MVFR, and VFR. It portrays general surface weather in a graphic form. The *radar summary chart* provides a summary of NEXRAD radar data. *Constant-pressure charts* analyze various layers of the atmosphere. Those usually used for aviation purposes depict layers at approximately 5000, 10,000, 18,000, 30,000, and 39,000 ft. The last set of charts displays observed atmospheric stability in the form of the lifted index, moisture available for precipitation in the perceptible water chart, and freezing level and average relative humidity.

These charts are available at flight service stations (FSSs) and through most commercial vendors with graphics capability, although the actual data displayed may vary. For example, most commercial surface charts do not contain station model data. Charts are also available from the National Weather Service (NWS) on the Internet at *www.weather.noaa.gov/fax/nwsfax.shtml.* The charts used in this book were obtained from this source.

Surface Analysis and Weather Depiction Charts

Sea-level pressure is the key element of the surface analysis chart. Observed station pressure, converted to sea level, allows analysis from a common reference. The data are computer analyzed for *isobars*— lines of equal sea-level pressure. As well as pressure patterns and fronts, the chart provides a look at surface wind flow, temperature, and moisture patterns, providing a primary source for the synopsis. The chart also serves to locate the position of vertical motion in the atmosphere, at and near the surface.

The surface analysis chart is prepared and transmitted every 3 hours. Observed data must be plotted and analyzed, so the chart is always old, sometimes several hours, by the time it becomes available. The chart always should be updated with current reports.

Most pilots have forgotten how to read station models. Some commercial weather vendors provide this information. Pilots do not need to decode the entire model, just the details significant to aviation. However, it is not within the scope of this book to decode these data. The interpretation of the various weather elements contained in the station models is the same as discussed in Chapter 1 in the cloud and precipitation sections and in this chapter under METAR/SPECI.

Refer to Fig. 3-3, the 1500Z, February 24, 1998, surface analysis chart. Convergence occurs along curved isobars surrounding a low or trough. Maximum convergence takes place at the low center or along the trough line. The figure shows surface winds blowing into the low in southern Nevada. Divergence—downward vertical motion— stabilizes the atmosphere, which decreases relative humidity and clouds. The figure also shows surface winds blowing out of the high in Alabama.

Note the anticyclonic outward flow from the high centers over the Midwest and Gulf Coast in Fig. 3-3. Maximum divergence takes place

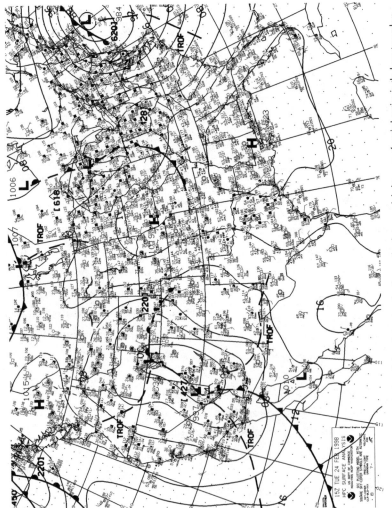

Fig. 3-3. The surface analysis chart displays the location of surface weather features and depicts areas of vertical motion in the lower troposphere.

at the high center. A high or ridge implies surface divergence. Troughs, with their cyclonic flow, produce upward vertical motion. This is occurring in the figure along the trough in southern California.

The significance of fronts was discussed in Chapter 1. Fronts run the spectrum from a complete lack of weather, to benign clouds that can be conquered by the novice instrument pilot, to fronts that spawn lines of severe thunderstorms and tornadoes that no pilot or aircraft can negotiate. Each front—for that matter, any weather system—must be evaluated separately, and then a flight decision can be made based on the latest weather reports and forecasts and the pilot's and airplane's capabilities and limitations.

One would certainly expect an icing potential associated with the fronts over the Great Lakes and New England, with surface temperature at or above freezing. However, the clear skies associated with the cold front in southern Indiana, Illinois, and Missouri present little, if any, icing potential. The cold front associated with the surface lows that runs through Utah, Nevada, and western Arizona would be another area for potential icing.

In Fig. 3-3 upslope is occurring over the plains in eastern Colorado and Wyoming associated with the low over Salt Lake City and the warm front. Temperatures ahead of the warm front are below freezing. Based on this information alone, a pilot could expect to pick up significant amounts of ice over Wyoming that would be carried all the way to the ground. Supercooled large droplets (SLDs) could be a factor as warm precipitation aloft falls into below-freezing temperatures below. However, remember that we are only evaluating this product alone.

The weather depiction chart is computer generated, analyzed, and transmitted every 3 hours as a record of observed surface data. Frontal positions are obtained from the previous surface analysis. The chart is broken into three categories—IFR, MVFR, and VFR—the same as the outlook portion of the area forecast. Information is hours old by the time the chart becomes available. The chart cannot consider terrain; nor is it intended to represent conditions between reporting locations. Gross errors between depicted categories and actual weather can occur. Since conditions can improve or deteriorate, data always must be updated with current reports. The weather depiction chart is not a substitute for current observations.

Station models on the weather depiction chart plot cloud height in hundreds of feet AGL, plotted beneath the model. Visibilities of 6 mi or less and present weather are entered to the left of the station. Sky cover and present weather symbols are the same as those used on the surface analysis chart.

Refer to Fig. 3-4, the 1600Z, February 24, 1998, weather depiction chart—produced 1 hour after the surface analysis chart in Fig. 3-3. The weather depiction chart provides a big, simplified picture of surface conditions. It alerts pilots to areas of potentially hazardous low ceilings and visibilities. The chart is often a good place to begin looking for an IFR alternate.

The warm front, discussed in the preceding section, is weak. Conditions ahead of the front are typically VFR to MVFR, although snow is falling in western Wyoming. This would tend to discount any severe icing potential in the region. Weak upslope is also indicated. Cloud layers are VFR, lowering to MVFR, with some IFR conditions as the moisture flows toward the continental divide. The cold front in southern California is also weak, again tending to discount any serious icing potential. Considerable rain and snow are associated with the systems over the Great Lakes and New England. This certainly would indicate serious icing potential in these areas. Conditions over New Mexico and the Southeast indicated little, if any, potential for icing.

The chart could be used to determine likely locations for a suitable alternate for an IFR flight into New England. An alternate to the south,

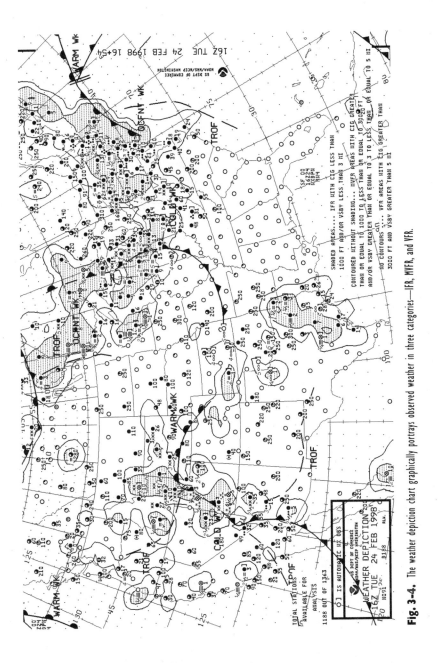

Fig. 3-4. The weather depiction chart graphically portrays observed weather in three categories—IFR, MVFR, and VFR.

Maryland or Virginia, would be indicated. Michigan has good weather, but be careful, with a cold front to the west, conditions could change rapidly.

I hope you can see how the analysis has changed significantly with only a look at two charts. For the "complete picture," all available charts and reports need to be analyzed.

Radar Summary Chart

The radar summary chart graphically displays a computer-generated summation of digital NEXRAD data. The date and time of the observation—time is important because the transmission system may make the report several hours old—appear on the chart. Charts contain information on precipitation type, intensity, configuration, coverage, tops and bases, and movement. Line and area movement, echo movement, and tops are depicted using standard symbols.

Pilots can expect to find holes in what the radar summary chart portrays as an area of solid echoes. This apparent inconsistency results from several factors. Targets farther from the antenna may be smaller than depicted due to range and beam resolution. NWS weather radars, at a range of 200 mi, cannot distinguish between individual echoes less than 7 mi apart. A safe flight between severe thunderstorms requires 40 mi, so this provides adequate resolution to detect a safe corridor. Additionally, as little as 20 percent coverage of VIP level 1 requires the entire grid to be encoded, so holes also occur with isolated and scattered precipitation. On the radar summary chart, large areas may be enclosed by relatively isolated echoes. However, pilots are cautioned that when determining intensity levels from the radar summary chart, always use the maximum intensity.

On the other hand, FSS or commercially available NEXRAD observations are usually reported in real time. They are analyzed in the same manner as the radar summary chart. However, the same cautions apply. If you are looking at a live radar picture at home or in the office, this does not mean that you can expect these conditions during your flight. However, if an active cold front is moving through the area, the radar may be an excellent source to identify when the system has moved through the area. Like all weather products, radar data must be analyzed within the context of all other reports and forecasts—the "complete picture."

The radar summary chart provides general areas and movement of precipitation for planning purposes only. Chart notations, such as *OM* (out of service for maintenance) or *NA* (not available) must be considered. As mentioned previously—and something that cannot be overemphasized—the chart always must be used in conjunction with other charts, reports, and forecasts. Once airborne, inflight observations—visible or electronic—and real-time radar information from an FSS or Flight Watch must be used.

FACT
- Radar summary charts may be 2 hours old.
- FSS and NEXRAD reports provide up-to-the-minute information.

Refer to Fig. 3-5, the 1635Z, February 24, 1998, radar summary chart. This time frame coincides with both the surface analysis and weather depiction charts discussed previously. There is apparently considerable precipitation associated with the cold front in the Southwest—*apparently*, because most of the intensity is VIP level 1 and 2. Even with relatively high cloud bases, pilots could expect icing associated with this system. The weak warm front over Wyoming is producing very little precipitation; therefore, serious icing, except in the area of light rain, is not likely. No echoes in eastern Texas indicate shallow stratus clouds and fog but no major weather or convective activity. However, the storm over the Great Lakes and New England is producing quite a bit of precipitation. Certainly, significant icing could be expected in this region. No thunderstorm activity was reported on this day, at this time. This is not unusual for the time of year. But remember: Reports or forecasts of thunderstorms imply severe icing!

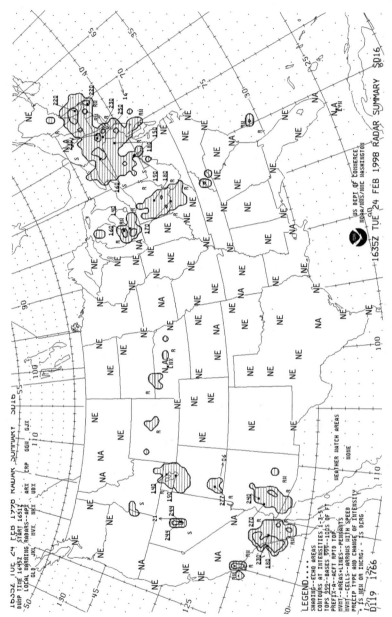

Fig. 3-5. The radar summary chart depicts precipitation as reported by NEXRAD Doppler weather radar.

Keep in mind that like thunderstorms and low-level wind shear, there are a few absolutes with icing, except for the hazard. We have only analyzed part of the picture, but I hope you can see how we are putting together the "complete picture."

Constant-Pressure Charts

Weather exists in two lower layers of the atmosphere—the troposphere and the stratosphere—and the boundary between them, the tropopause. Pilots fly and weather occurs in three dimensions, so a need exists to describe the atmosphere within this environment. The NWS prepares several constant-pressure charts. These computer-generated charts are transmitted twice daily based on 0000Z and 1200Z upper-air observations.

Each chart represents a constant pressure level, analyzed for altitude or height in meters above sea level. Lines, known as *contours,* connect areas of equal height. However, unlike the surface, wind blows parallel to the contours, due to the lack of friction; only pressure gradient and Coriolis forces are present.

Constant-pressure analysis charts depict wind direction and speed using standard symbols, except for the contraction *LV,* which indicates light and variable. Temperature is plotted in degrees Celsius. However, unlike surface station plots, the temperature–dew point spread, or depression, appears instead of the dew point temperature. Darkened station circles, plotted with a temperature–dew point spread of 5°C or less, indicate a moist atmosphere and an icing potential. An *X* indicates a temperature–dew point spread greater than 20°C. With temperatures less than −41°C, air is too dry to measure dew point, and it is omitted. Under these conditions, there is little, if any, icing potential.

The 850- and 700-mb charts represent the lower portion of the troposphere, approximately 5000 and 10,000 ft, respectively. The 850-mb chart may be more representative of surface conditions west of the Rockies than the surface analysis chart. In the West, areas of frictional convergence/divergence and upslope/downslope can be located on the 850-mb chart.

These charts are particularly useful in monitoring cold- and warm-air advection. When *isotherms*—dashed lines of equal temperature—cross contours at right angles, the temperature properties of the air mass are advected (moved) in the direction of the winds. At the 850- and 700-mb levels, warm-air advection produces upward motion and cold-air advection produces downward motion. Forecasters look for a warm tong, or warm nose, to indicate vertical motion in the lower troposphere and as a factor for potentially serious icing.

Areas of moisture can be determined by examining station models for temperature–dew point spread. Icing is implied in areas of visible moisture with temperatures between 0°C and −10 to −15°C.

Figure 3-6 shows the 700-mb chart at 1200Z on February 24, 1998. The 0°C isotherm extends from northern Baja California through southern Kansas and then to Georgia. The −10°C isotherm passes through central California to the low in southeast Oregon, to the Great Lakes, and then north into Canada. Typically, this would be the area for the most significant icing, in areas of clouds and precipitation, at the 10,000-ft level. In Fig. 3-6 isotherms parallel contours; on this day there is little warm- or cold-air advection at this level. Based on temperatures–dew point depressions, this layer of the atmosphere is relatively dry associated with the frontal system in the West but quite moist in the Northeast associated with that frontal system and low.

Probably the most important and useful chart—maybe even more important to meteorologists than the surface analysis chart—the 500-mb chart describes the atmosphere in the middle troposphere, which is an altitude of approximately 18,000 ft. This chart provides important pressure, wind flow, temperature, and moisture patterns and can be used to determine areas of vertical motion at this level.

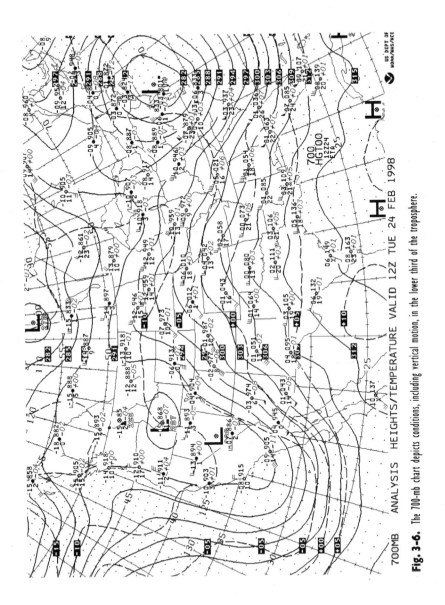

Fig. 3-6. The 700-mb chart depicts conditions, including vertical motion, in the lower third of the troposphere.

Unlike the surface analysis chart, where upward vertical motion takes place along a trough line, upward vertical motion at the 500-mb level takes place between the trough and ridge lines. Troughs transport cold air down from the north and warm air up from the south. Warm air rides northward on the east side of the trough—trough-to-ridge flow—so the air is lifted as it moves northward, producing upward vertical motion. When moisture is present, clouds and precipitation develop in the midtroposphere. Conversely, in ridge-to-trough flow, cold air sinks southward, producing downward vertical motion with typically clear, dry conditions in the midtroposphere. Clouds and precipitation frequently accompany upper-level lows and troughs, even without surface frontal or storm systems.

Cold-air advection destabilizes the atmosphere above the 500-mb level, and warm-air advection stabilizes the atmosphere. This is opposite to the effects of cold- and warm-air advection near the surface. Rising air will be warmer than surrounding air, so cold-air advection enhances thunderstorms by promoting vertical development, sometimes referred to as a *cold low aloft*. Cold lows aloft tend to move slowly and erratically. Warm-air advection at this level strengthens high-pressure ridges and diminishes low-pressure troughs. Moisture at the 500-mb level can be determined by darkened station models. This chart is a good indicator of high-level icing in summer months and with storms that are either well developed or contain tropical moisture.

Figure 3-7 shows the 500-mb chart at 1200Z on February 24, 1998. Temperatures are all below $-10°C$, except for extreme southern Texas and southern Florida. Therefore, we would not expect much in the way of serious icing above the 18,000-ft level for the United States on this day. Station plots show quite a bit of moisture in the trough-to-ridge flow from southern California to the Midwest and mostly dry conditions in central portions of the Midwest and southeastern United States in the ridge-to-trough flow. Moisture again is prevalent in the Northeast, associated with the upper-level low. Just exactly what we would expect.

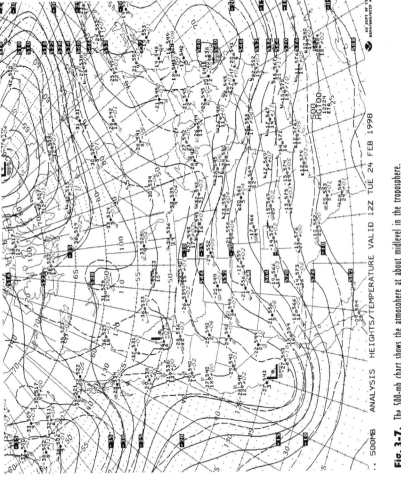

Fig. 3-7. The 500-mb chart shows the atmosphere at about midlevel in the troposphere.

From the discussion so far, one would expect icing potential from about the 10,000-ft level through about 20,000 ft in the area of precipitation over southern California and Arizona. The icing potential over New England would be at lower altitudes due to colder air in that region. However, one could expect the possibility of more serious icing resulting from the overrunning warmer air from the occluded and warm fronts offshore. (Recall the likely locations of SLD icing from Chapter 1, Fig. 1-6.) One would expect icing from the surface to about the 15,000-ft level in this area.

The 300- and 200-mb charts provide details of pressure, wind flow, and temperature patterns at the top of the troposphere and occasionally into the lower stratosphere. These charts indicate the strength of features in the lower atmosphere. Strong storm systems on the surface are reflected in the 300- and 200-mb patterns, whereas weaker systems lose their identity at these levels. At middle latitudes, such as the United States, the jet stream usually can be found on the 300- and 200-mb charts.

Figure 3-8 shows the 300-mb chart at 1200Z on February 24, 1998. The −40°C isotherm is south of most of the United States. Therefore, especially without convective activity, one would not expect any significant icing at or above 30,000 ft. With the exception of the northern Rockies and the upper-level low in the Northeast, the air at this level is dry. If we compare the radar summary chart (Fig. 3-5), precipitation tops throughout the country are all below this level.

Moisture/Stability Charts

The *moisture/stability series*—formally the *composite moisture-stability chart*—consists of four charts: the lifted index analysis, precipitable water, freezing level, and average surface to 500-mb relative humidity charts. Available twice daily, these charts are computer generated from upper-air observation data. Because of computation and transmission times, these charts are about $4^{1}/_{2}$ hours old by the time they become available.

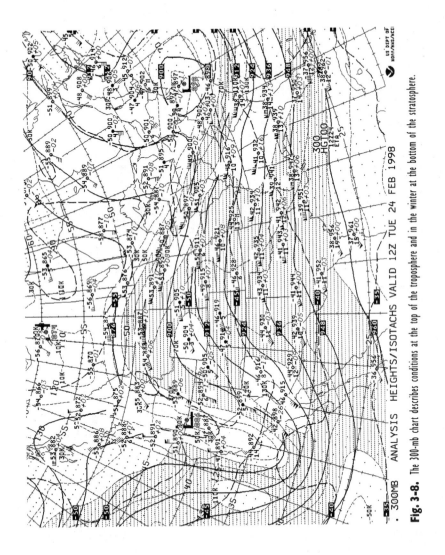

Fig. 3-8. The 300-mb chart describes conditions at the top of the troposphere and in the winter at the bottom of the stratosphere.

The *lifted index* provides an indication of atmospheric moisture and stability—thunderstorm potential at the time of observation. The lifted index compares the temperature a parcel of air near the surface would have if lifted to the 500-mb level and cooled adiabatically, with the observed temperature at 500 mb. The index indicates stability at the 500-mb level. The index can range from +20 to −20, but generally it remains between +10 and −10 (the figures are strictly an index, not a representation of temperature). A positive index indicates a stable condition; high positive values indicate very stable air. A zero index indicates neutral stability. Values from 0 to −4 indicate areas of potential convection. Large negative values from −5 to −8 represent very unstable air, which could result in severe thunderstorms—should convection develop.

The *K index* evaluates moisture and temperature. The higher the K index, the greater is the potential for an unstable lapse rate and the availability of moisture. The K index must be used with caution; it is not a true stability index. Large K indexes indicate favorable conditions for air-mass thunderstorms during the thunderstorm season. K index values can change significantly over short periods as a result of temperature and moisture advection.

The lifted index (LI) appears above the K index in the plotted data (Fig. 3-9). *Isopleths*—lines equal in number or quantity of stability— are plotted beginning at zero and then for every four units (plus and minus). Slightly unstable conditions exist in northern California, southern Texas, Florida, and along the southern Atlantic Coast states. Large K indexes occur along the southern Atlantic Coast states. The chart therefore would indicate a potential for thunderstorms in the southeastern United States. Conversely, high positive lifted indexes and small K values are observed over the rest of the country, indicating little moisture and a stable lapse rate.

This chart can be used to locate areas of significant instability (LI<0) and areas of high moisture content in the lower portion of the troposphere (K index>25 to 30). Negative lift indexes and large K

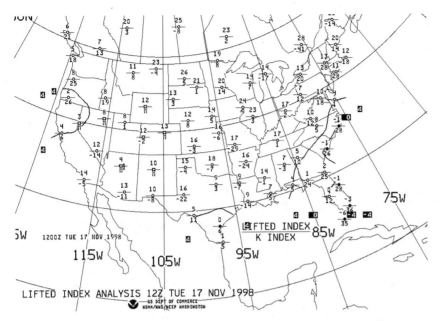

Fig. 3-9. The lifted index provides an indication of atmospheric moisture and stability—thunderstorm potential.

values do not necessarily mean that thunderstorms or serious icing will develop; they are merely one indicator.

The *precipitable water chart* (Fig. 3-10) analyzes water vapor content from the surface to 500 mbar. The chart indicates the amount of liquid precipitation that would result if all the water vapor were condensed. Darkened station circles indicate large amounts of available water. Isopleths of precipitable water are drawn at $1/4$-in intervals. The panel is more useful for meteorologists concerned with flash floods. The chart can be used to determine if the air is drying out or increasing in moisture with time by looking at the wind field upstream from a station. Therefore, a pilot can get an excellent indication of changes in moisture content. For example, considerable moisture exists over extreme southern Texas, Florida, and the southeastern Atlantic Coast. With the stability obtained from the lifted index, these areas would have the greatest thunderstorm potential.

In general, for this day, the atmosphere over the United States is relatively dry, as indicated by the precipitable water analysis. Serious

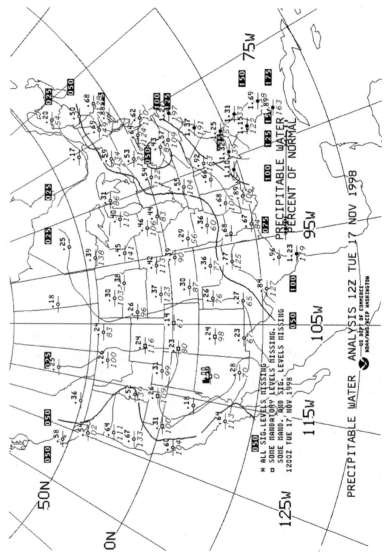

Fig. 3-10. The perceptible water chart analyzes water vapor content between the surface and the 500-mb level.

icing potential and possible SLDs would be associated with areas of high precipitable water.

By applying wind flow from observed and forecast charts, one can obtain a general sense of movement. For example, if the Bermuda high dominated the western Atlantic one would expect the moisture and instability of the southeastern United States to be advected into the Mid-Atlantic coastal states. Should a weak cold front be moving through southern Texas, one would conclude that dryer, stable air would be entering the region.

The *freezing-level panel* (left half of Fig. 3-11) plots the lowest observed freezing level. Multiple entries indicate inversions and areas of above-freezing temperatures aloft. This has occurred in central Wyoming. Notice that the surface is below freezing (BF). Above-freezing air exists aloft between 5700 and 10,600 ft MSL. Implications are that a pilot could find above-freezing temperatures below 10,600 ft and above 5700 ft. Should precipitation be falling, freezing rain, freezing drizzle, or ice pellets could occur from 5700 ft to the surface. Keep in mind that these are observed data and must be used with forecast information contained in the AIRMET bulletin, SIGMETs, and other forecasts for flight planning.

From Fig. 3-11 it is apparent that freezing levels are high for the southern third of the nation, with freezing levels in the 13,000- to 15,000-ft range. This implies no icing below these levels. However, with moisture above the freezing level, icing could be possible into the lower flight levels.

Two crossings can occur with surface temperature below freezing. For example, at the station in western Montana, temperatures are below freezing from the surface to 3900 ft MSL, with a layer above freezing air from 3900 to 6800 ft.

Figure 3-12 illustrates three crossings of the freezing level over the station in central Illinois. Note first that the surface temperature is

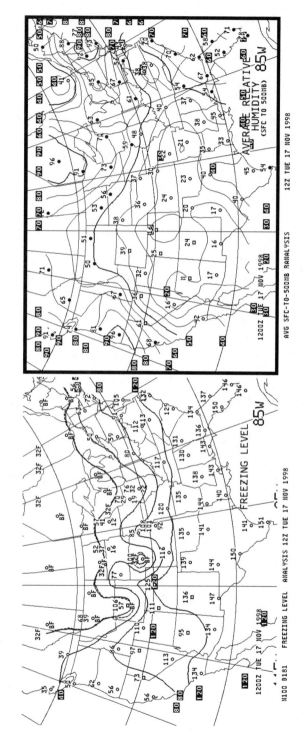

Fig. 3-11. The freezing-level panel plots the lowest observed freezing level; the average relative humidity panel indicates large-scale moisture content in the lower part of the troposphere.

above freezing. The below-freezing (32°F) isopleth is north of the station. The first crossing occurs at 2200 ft MSL (22 below the station circle), the second crossing occurs at 3200 ft MSL (32), and the third at 7600 ft MSL (76). In this example, air temperature that is above freezing occurs from the surface to 1900 ft and again between 3200 and 7600 ft—the gray-shaded area to the right of the 0°C isotherm. This could be significant should an altitude be required to shed ice. A word of caution: Do not bet your life that these conditions will exist. Hoping that an inversion aloft will be available should never be your only escape plan.

CASE STUDY

The pilot of a Beechcraft Sierra (BE24) received a preflight weather briefing and was advised of occasional moderate mixed and rime icing, as well as isolated severe mixed and clear icing, along the route of flight. Enroute the pilot encountered icing conditions, and ice accumulated on the airplane. The pilot reported losing power and subsequently made a forced landing in a residential area. The pilot stated: "At 5000 ft, I expected to be above the freezing level, due to a temperature inversion. This was the case until I reached the Philadelphia area; I then encountered light mixed icing."

The probable cause, determined by the National Transportation Safety Board (NTSB), was the pilot's improper planning/decision making and resulting flight into known icing conditions. Under such circumstances, the pilot's only safe option was to reverse course to an area of warmer air. This action may have required the declaration of an emergency and its possible repercussions. However, any repercussions would have been infinitely preferable to the actual outcome.

Freezing-level data—RADAT—in the past appeared in remarks on 00Z and 12Z surface observations associated with radiosonde (upper air) observation sites. The RADAT has been removed from these observations in the contiguous United States. At present, RADAT information is only available from an FSS. In the future, direct user access terminals (DUATs) or the Internet may make these data

MULTIPLE FREEZING LEVELS

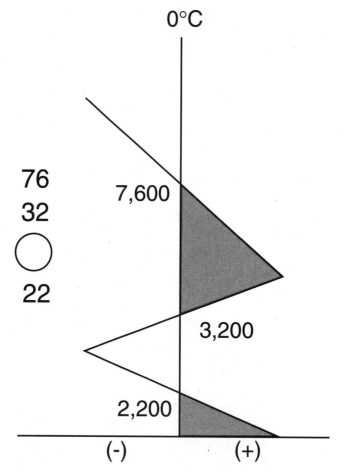

Fig. 3-12. Multiple freezing levels indicate inversions aloft, with the possibility of freezing precipitation and the presence of supercooled large droplets.

available. Radiosonde release sites can be found in the back of the appropriate *Airport/Facility Directory.*

The data consist of the freezing level in hundreds of feet MSL and relative humidity at that level. For example, in .../RADAT 32078, the first two digits represent relative humidity (32 percent) at the freezing level of 7800 ft MSL. With multiple freezing levels, the highest relative humidity is indicated by the letter *L,* meaning lowest; *M,* middle; or *H,* highest crossing of the 0°C temperature. For example, the RADAT for the station in Fig. 3-12 might appear …/*RADAT98M022032076*. In this example, the highest relative humidity is 98 percent, which occurred at the middle (*M*) 3200-ft crossing. High relative humidity indicates moisture and the possibility of icing at and above the freezing level. This is similar to the information available on the freezing-level chart, except that relative humidity is not plotted on that chart.

Other RADAT and chart symbols indicate the sounding was missed (*M*), due to equipment trouble or other factors, or delayed (*NIL*). If the entire sounding is below 0°C, RADAT *ZERO* or *BF* will be reported.

The *average relative humidity panel* (right half of Fig. 3-11) analyzes the average relative humidity from the surface to 500 mb. Darkened station plots indicate high humidity—50 percent or greater. *Isohumes*—lines of equal relative humidity—are drawn at 10 percent intervals. The chart indicates large-scale moisture content in the lower part of the troposphere. Frequently, clouds and precipitation are indicated by relative humidities of 50 percent or greater. These conditions exist in the extreme Northwest and along the Atlantic Coast. Note that the humidity over central Nebraska is relatively—I know, I did it again—dry, thus indicating the absence of clouds, precipitation, and serious icing.

These charts are useful in determining the characteristics of a particular weather system in terms of stability, moisture, and possible aviation weather hazards. Even though these charts are hours old by the time they become available, the weather systems will tend to move these characteristics with them, although caution should be exercised because characteristics may be modified through development, dissipation, or movement of weather systems.

Satellite Images

There are two main types of meteorologic satellites: polar orbiters and equator orbiters. Pilots usually have access to the equator orbiters or the Geostationary Operational Environmental Satellites (GOESs). This discussion will be limited to GOESs. GOESs orbit directly over the equator at approximately 19,000 nautical miles. They make one revolution every 24 hours. From the satellite's view, the earth appears to remain stationary—thus the name *geostationary*. Satellite images are available on many Internet sites. Those used here are available from the NWS at *www.goes.noaa.gov*.

Satellite interpretation is a science in itself. Therefore, the discussion will be limited to two basic images: visible and infrared. *Visible* images, as the name implies, are a "snapshot" of conditions on earth. Resolution of visible images ranges between $1/2$ and 2 nautical miles. *Infrared* (IR) images are temperature pictures. That is, the satellite senses the temperature of an area with a resolution of approximately 5 nautical miles.

Various types of clouds and terrain reflect different amounts of sunlight. The best reflectors are large cumulonimbus clouds. Thin clouds or areas of very small clouds appear darker because less sunlight is reflected. Below is a list of the reflectivity of various surfaces, beginning with the brightest.

- Large thunderstorms
- Fresh new snow
- Thick cirrostratus clouds
- Thick stratocumulus clouds
- Snow 3 to 7 days old
- Thin stratus clouds
- Thin cirrostratus clouds
- Sand
- Sand and brushwood
- Forest
- Water surfaces

IR images begin by portraying different temperatures as black, shades of gray, and white. Black is the warmest and white the coldest. Typically, black represents a temperature of about 33°C and white about −65°C, with the gray shades representing decreasing temperature toward the white end. In fact, there are 256 distinct shades from black to white. Basically, though, warm temperatures are dark, cool temperatures are gray, and cold temperatures are light.

Computer technology allows the enhancement of IR images. This technique allows the operator to highlight areas of interest. Colors can be assigned to specific values within the 256 shades. This results in the variety of color satellite images seen on television and available on various Internet sites. However, without knowing the exact "enhancement curve," specific interpretation is difficult.

Like chart interpretation, best results are obtained from comparing the visible image with the same-time-frame IR image. Often what may be misinterpreted or ambiguous on one image can be resolved by comparing it with the other image. Unfortunately, this does not work at night, when only the IR image is available. (I had a pilot ask me for the visible satellite image one morning at about 6 A.M. In a feeble attempt at humor, I replied, "It's not available, the flash bulb is burnt out." The pilot replied, "Yes, you've been having a lot of trouble with that satellite lately." Oh well.) And again, like chart interpretation the best picture—if you will—is obtained by a comparison of chart and satellite information.

Now if you have the capability to view several images in succession—a satellite loop—you can frequently get a sense of weather movement, development, and dissipation. Viewing a loop also will help distinguish between snow cover and cloud cover.

Well, I have just condensed several days of classroom study into one page. Please remember that these are only the very basics.

Figures 3-13 through 3-16 are four satellite images. Figures 3-13 and 3-14 are from GOES-W (west), located at about 135° of longitude.

Figures 3-15 and 3-16 are from GOES-E (east), located at about 75° of longitude. Again, there are visible and IR images, respectively. They were taken on February 24, 1998, between 1815Z and 1830Z. They represent approximately the same time period as the surface analysis chart (see Fig. 3-3), weather depiction chart (see Fig. 3-4), radar summary chart (see Fig. 3-5), and 500-mb chart (see Fig. 3-7) presented earlier.

Let us begin with Fig. 3-13. Its main feature is the frontal system depicted on the surface analysis chart. It appears that there is a rapidly moving cold front, since the satellite shows the frontal boundary already moving through central Arizona. There is an extensive cloud shield associated with the low over northern Utah and extending through Idaho, Montana, and the Dakotas. Tops are highest, associated with the front, over Arizona—represented by the bright portion of the image. Note what appears to be a low cloud—relatively dark—over northern Texas.

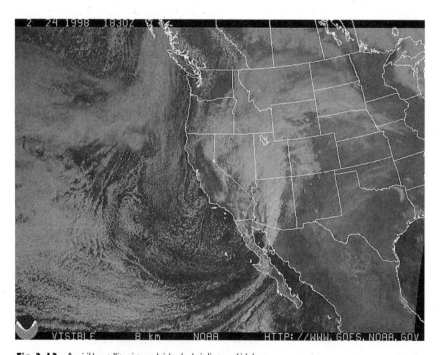

Fig. 3-13. On visible satellite pictures, bright clouds indicate a thick layer.

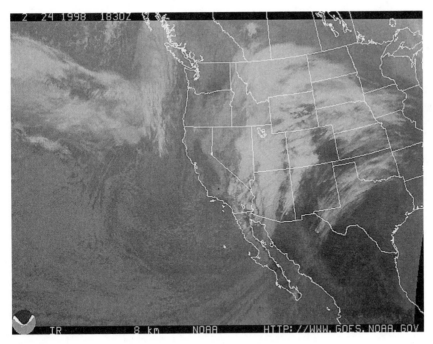

Fig. 3-14. On infrared satellite pictures, bright clouds indicate cold, high tops.

Compare the cloud cover over southern California with the radar summary chart in Fig. 3-5. The satellite image confirms that the area of precipitation is more scattered than depicted on the radar chart.

Now let us compare the IR image in Fig. 3-14. The brightness of the cold front confirms cold, thus high, tops. However, note how the frontal clouds are markedly dark for that portion of the front over the Pacific Ocean. In this area, tops are quite low, and the front is weak and diffuse. Tops are much lower over western Nevada and northern and central California. High tops are also associated with the cloud shield. This is certainly consistent with the surface analysis, weather depiction, and radar summary charts. It also confirms less coverage of precipitation in southern California than indicated by the radar summary chart. Now look at northern and western Texas. What appeared in the visible image to be low clouds is revealed to have very cold tops. In fact, this is a thin band of cirrus clouds. This signature

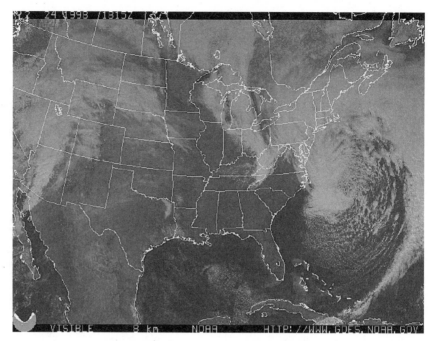

Fig. 3-15. On visible images, thin clouds appear relatively dark, like those over the Texas panhandle.

typically represents the northern edge of the subtropical jet stream, and the clouds are called *jet stream cirrus clouds.*

The cloud patterns in the visible image in Fig. 3-15 are very consistent with the information depicted on the surface analysis, weather depiction, and radar charts. Note the considerable upslope on the western slopes of the Appalachian Mountains in West Virginia, Virginia, and eastern Tennessee. Off the East Coast is a classic "comma cloud" from a well-developed storm system. Finally, note the cloud pattern in eastern Texas.

Now look at the IR image in Fig. 3-16. High tops are associated with the storm system over New England, and relatively low tops are seen over the Great Lakes and southern Appalachians. This is again consistent with the surface analysis, weather depiction, and radar summary charts. The southeast United States is clear, consistent with the high pressure over that area.

These satellite images are also consistent with the 500-mb constant-pressure chart: clouds and precipitation in the trough-to-ridge flow over the western United States and clearing conditions in the ridge-to-trough flow over the Midwest.

The satellite often can confirm or refute areas of weather depicted on various weather charts and predicted in forecasts products. However, as with every other product, its limitations must be understood, and the information must be taken along with all other available data. This is especially true at night when only the IR image is available.

Notices to Airmen (NOTAMs)

Failure to check NOTAMs has led many pilots into embarrassing and potentially hazardous situations. Increased use of direct user access

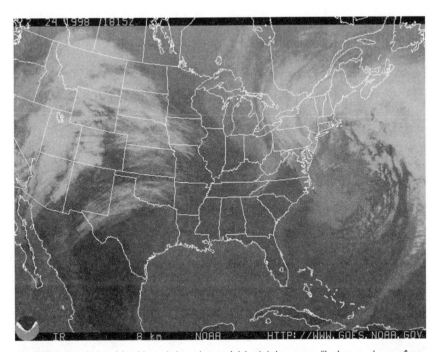

Fig. 3-16. Low clouds on infrared images look gray because of their relatively warm tops, like those over the upper Great Lakes and off the Carolina coast.

SOLUTIONS TO TRY THIS CHALLENGE

I asked some questions during the discussion of charts. Recall that the weather depiction chart shows extensive areas of MVFR and IFR in southeastern Texas. The surface analysis chart shows a southeasterly flow at the surface along the Texas Gulf Coast. The Gulf of Mexico is an abundant source of moisture. From a knowledge of terrain, we know that elevations increase toward the Northwest. The satellite images confirm that these are low clouds—dark gray, warm tops on the IR image. These clouds are indeed low stratus clouds and fog, cause by the anticyclonic flow around the high over the Southeast and upslope. The visible image (Fig. 3-15) shows that the area has almost dissipated along the Gulf Coast, although clouds remain in northeast Texas. The stratus cloud is not as extensive as depicted on the weather depiction chart, a common anomaly of this product. There is certainly no icing potential in this area.

terminals (DUATs) and other commercially available briefing systems may mean that interpreting and understanding NOTAMs will take on even greater significance. Add to this recent changes in the FAA's NOTAM system with the introduction of ICAO contractions and date/time groups. These factors will challenge pilots to a much greater degree than in the past.

The FAA advertises the status of components or hazards in the National Airspace System (NAS) through aeronautical charts, the *Airport/Facility Directory,* other publications, and the National Notice to Airmen System. Changes are normally published on charts, in the directory, or in the *Notices to Airmen* publication. The need for current charts and publications cannot be overemphasized.

Aeronautical information not received in time for publication is distributed on the FAA's telecommunications systems, including unanticipated or temporary changes or hazards when the duration is for a short period or until published. Unpublished NOTAMs are not necessarily given during FSS abbreviated or outlook briefings but are provided routinely as part of a standard briefing.

NOTAMs consist of NOTAM (L), NOTAM (D), FDC NOTAMs, and those published in the *Notice to Airmen* publication. Only distributed locally, NOTAM (L)s advertise conditions or hazards that do not meet the criteria for a NOTAM (D). Over the years, fewer and fewer items remain in this category. NOTAM (L)s are essentially "nice to know" information. FDC NOTAMs contain regulatory information (chart amendments, changes to instrument approach procedures, and temporary flight restrictions). The *Notice to Airmen* publication contains information known far enough in advance to be published and mailed.

NOTAM (D)s contain information that might influence a pilot's decision to make a flight or require alternate routes, approaches, or airports. They are considered "need to know" and are issued for certain landing area restrictions, lighting aids, special data, and air navigation aids that are part of the National Airspace System. NOTAM (D)'s are issued for public-use airports listed in the *Airport/Facility Directory.*

Terminology and Format

Like METAR, PIREPs, and TAF, NOTAM codes were changed in 1996. However, for NOTAMs, the change was postponed several times. (It appears that the FAA's philosophy is: "We never have time to do it right, but we always have time to do it over.") There were two major changes: the use of international contractions and a ten-digit date/time group.

The date/time group consists of year, month, day, and time (UTC). Times used on NOTAMs are now all UTC. The day begins at 0000Z and ends at 2359Z. For example, *9810291400* decodes as year *98* (1998), month *10* (October), day *29*, and time 1400Z. A common contraction used with date/time groups is *WEF,* which translates as "with effect from or effective from." Use care when determining effective times! These date/time groups can be very confusing. The absence of a date/time group means the condition is in effect and will continue until further notice (*UFN*). However, *UFN* in the NOTAM text is not transmitted.

Runways are identified by magnetic bearing (12/30, 12, or 30). If magnetic bearing has not been established, the runway is identified by the nearest eight points of the compass (*NE-SW, N 200 N-S RY*).

NOTAM (D)s begin with the character *!*, which is an automatic data-processing code. This is followed by the accountability LOCID, e.g., *OAK* for Oakland, California. Next comes the NOTAM number. NOTAM (D)s are numbered by month of issuance and are numbered consecutively during the month. NOTAM number 03/005 was issued during March and indicates the fifth NOTAM issued in that month. Next comes the specific facility or location. This is the LOCID of the airport, facility, or navigational aid affected. The last two items are the condition reported and effective times, if required.

> An example of NOTAM format:
>
> !OAK 03/005 OAK VORTAC OTS WEF 9803021800-9803022200
> This is Oakland, California, NOTAM 03/005. It was issued during the third month of the year (*03*/005) and was the fifth NOTAM for Oakland (03/*005*). The Oakland VORTAC is scheduled to be out of service (OTS) on March 2, 1998 (*980302* 1800) at 1800Z until March 2, 1998 at 2200Z.

NOTAM (D)

For our purposes, this discussion will be limited to icing conditions that restrict landing area operations. These NOTAM (D)s are issued for restrictions to landing areas (runways and waterways), including closures, braking action, and problems due to snow, ice, slush, or water. A complete discussion of NOTAMs would take up a whole chapter—in fact, it does in *The Pilot's Guide to Weather Reports, Forecasts, and Flight Planning.*

Airport management personnel are responsible for reporting restrictions for NOTAM issuance. FAA Flight Service Station controllers format and disseminate the information. The following icing information receives NOTAM (D) distribution:

- Runway friction measurements

- Runway breaking action

- Snow

- Ice

- Slush

- Water conditions

Civil pilots now have access to objective runway friction measurements at some airports. The Greek letter mu (μ), pronounced "myôô," designates a friction value representing runway surface conditions. Values range from 1 to 100, where 0 is the lowest friction value and 100 is the maximum. With snow or ice on the runway, a mu value of 40 or less is the level where the aircraft braking performance starts to deteriorate and directional control begins to be less responsive. The lower the mu value, the less effective braking performance becomes and the more difficult direction control becomes.

Airport management conducts friction measurements on runways covered with snow or ice. NOTAMs are issued only when one or more values are below 40. Values are reported for the first third, middle third, and last third of the runway. Pilots can expect to receive mu values from ATC or as a NOTAM (D). When friction measurements are reported for more than one runway, each will be advertised as a separate NOTAM specifying the runway. For example:

!DCA 12/005 DCA 36 MU 42/35/48

At Washington's Ronald Reagan Airport (DCA), the runway 36 mu values are 42, 35, and 48 for the first, middle, and last thirds of the runway, respectively.

No correlation has been established between mu values and the descriptive terms *FAIR, POOR,* and *NIL* used in braking-action reports. Pilots should use mu information with other knowledge, including type, weight, wind conditions, and previous experience.

When braking action is reported by airport management, it receives NOTAM (D) dissemination. Braking action reported by a pilot is distributed as a PIREP. In both cases, braking action (BA) is reported as *FAIR, POOR,* or *NIL.* The type of vehicle, when used, will not appear in the NOTAM.

There appears to be no hard and fast definitions of reported braking action. *FAIR* would tend to indicate that breaking action was certainly not what could be expected on a dry runway but the pilot was able to control the airplane. *NIL* represents a lack of any breaking action. The pilot has essentially no breaking control. And *POOR* is somewhere in between. Any of these reports or a mu NOTAM should alert the pilot to breaking problems at the airport. The bottom line: If you have not had training or experience with snow- or ice-covered runways, avoid them!

Snow, ice, and water are self-explanatory. *Slush* is a soft, watery mixture of snow or ice on the ground that has been reduced by rain, above-freezing temperatures, or chemical treatment. When snow, ice, slush, and water conditions are reported, their depth is expressed in terms of thin—less than $1/2$ in. (*THN*), $1/2$ in., and 1 in. Additional amounts are reported in 1-in. increments. If the surface is not completely covered, it is reported as having patches (*PTCHY*). The absence of *PTCHY* means the entire landing area is covered.

Reports of snow (*SN*) on the runway may be modified by one of the following:

- *LSR*, loose snow
- *PSR*, packed or compacted snow
- *WSR*, wet snow (not slush)
- *RUF FRZN SN*, rough frozen snow

Conditions may be combined to better describe the surface. For example, *THN LSR OVER 1 IN PSR* translates to less than $1/2$ in loose snow and over 1 in. of packed snow.

!CLM 12/008 CLM 8/26 THN SN

In this example, the Port Angeles, Washington, runway 8/26 is reported covered with thin snow. *THN* indicates that the layer is less than ¹/₂ in. deep.

Ice is reported using the contraction *IR* or *SIR*. *SIR* means packed or compacted snow and ice on the runway. Other descriptors may be used to better define conditions.

!PAE 02/001 PAE PTCHY 1 IN RUF IR

There is patchy 1 in. of rough ice on the runway at Paine, Washington.

Slush is reported using *SLR*. For example:

!IAD 01/023 IAD 1L/19R 1/2 IN FRZN SLR

At Washington's Dulles International Airport there is ¹/₂ in. of frozen slush (may be described as *RUF IR*, but not wet snow).

Water on the runway is reported in a similar manner, using the contractions *WTR*. It may be reported as *PTCHY* but never as puddles.

!EWB 01/025 EWB PTCHY THN WTR

At New Bedford, Massachusetts, there are patchy areas of thin—less than ¹/₂ in.—water covering the airport surfaces.

Drifting or drifted snow (*DRFT*) describes one or more drifts. When drifts are variable in depth, the greatest depth is reported. When drifts are identified, pilots should consider that these conditions prevail throughout the airport surface—runways, taxiways, and ramps.

!SFF 01/020 SFF 4 IN LSR 9 IN DRFT

In this example, 4 in. of loose snow covers all airport surfaces at Spokane's Felts Field, with 9-in. snow drifts.

Plowed (*PLW*) or swept (*SWEPT*) indicates that a portion of the surface has been cleared. (*PLW* will not be reported when the entire runway has been plowed.) It is either bare or has depth, coverage, and conditions different from the surrounding area, which appear in the NOTAM. When known, the surrounding areas will be specified as remainder (*RMNDR*).

!MHT 01/067 MHT 17/35 SWEPT 100 WIDE RMNDR THN SIR

Runway 17/35 has been swept 100 ft wide. The remainder of the runway at Manchester, New Hampshire, is covered with thin snow and ice—less than $1/2$ in.

Treated runways are also reported. A sanded runway (*SA*) means that the entire runway has been sanded. If less than the published dimensions have been treated, it is indicated in the NOTAM. Since a deicing liquid (*DEICED LIQUID*) or solid (*DEICED SOLID*) may be operational, it is reported in the NOTAM.

!PSM 01/054 PSM 16/34 PTCHY THN IR SA 100 WIDE DEICED LIQUID/SOLID 100 WIDE

At Pease International, Portsmouth, New Hampshire, runway 16/34 has patchy, thin—less than $1/2$ in.—ice on the runway. The center 100 ft has been sanded, and both liquid and solid deicer has been used.

Snowbanks (*SNBNK*) or berm (*BERM*) is assumed to be at the runway edge. Should the runway be plowed or swept, the *SNBNK* is at the edge of plowed or swept area.

!BTV 01/014 0B5 PTCHY IR 3FT SNBNK BA NIL

Turners Falls Airport, Montague, Massachusetts, reports patchy ice on the runway, 3-ft snowbanks, and braking action nil.

Using the NOTAM System

As part of the standard briefing from an FSS, pilots receive any pertinent NOTAMs that are on hand. Pilots can expect to receive NOTAMs on the status of NAVAIDs, airway changes, and airspace restrictions, unless there is a temporary NOTAM system outage. In such a case, pilots will be advised of this fact and have to check with FSSs en route and at the destination to ensure receipt of current NOTAMs. This should include all pertinent landing area information. Information contained in the *Notices to Airmen* publication will only be provided on request.

THE BOTTOM LINE

Here's the bottom line: Request a standard briefing. NOTAMs are not necessarily provided during abbreviated or outlook briefings. If the specialist does not mention NOTAMs, ask. Briefers are human, and NOTAMs are easy to overlook. Keep in mind that NOTAMs are not normally available on FAA mass dissemination systems (TWEB, TIBS, or PATWAS). A check of the *Airport/Facility Directory* should be a standard part of flight planning; significant information may be published. If you want, request local NOTAMs from the tie-in FSS prior to descent and landing. Finally, if you do not have access to the *Notices to Airmen* publication, ask the briefer to check.

Pilots using DUAT or other commercially available systems will have to decode and translate NOTAMs. Remember the DUAT disclaimer: "Nonassociated FDC NOTAMs are available. Do you request them?" These would include temporary flight restrictions and airway changes. These systems do not provide NOTAM (L)s, GPS NOTAMs, or information contained in the *Airport/Facility Directory* or the *Notice to Airmen* publication. The contents of these documents still remain the responsibility of the pilot. Should any doubt exist about the meaning or intent of a NOTAM, consult an FSS for clarification.

Checking NOTAMs is like going to the restroom before a flight. You know you should, but sometimes it's just not convenient. Either oversight can lead to a very uncomfortable flight!

ICING MYTH

Pilot reports of icing are always overestimated.

CHAPTER 4

Icing Forecasts

cing is as difficult to forecast as it is hazardous. Forecasters must determine which areas contain enough moisture to form clouds, which cloud areas will most likely contain supercooled droplets during the forecast period (6 hours, with an additional 6-hour outlook), and the freezing level. Virtually all ice is found in about 5 percent of clouds. This is due in part to the cyclic development of clouds, as discussed in Chapter 1. This helps explain the transitory nature of this phenomenon and the perceived inconsistency of pilot icing reports. One pilot can fly through an area with no serious icing, and $1/2$ hour later another pilot's aircraft falls out of the sky.

Needless to say, forecasting icing is not an easy task. Pilots should consider these reports as forecasts of icing potential. They alert the pilot to the need to consider the possibility of icing in clouds and precipitation within the areas and altitudes specified.

Chapter 1 delineated the atmospheric properties that create aircraft icing. Hopefully, readers gained an appreciation of the complexity of this phenomenon. Chapter 3 pointed out that the only direct observations of aircraft icing are by the pilot, and if such observations are reported at all, PIREPs tend to be very subjective. (Progress is being made to equip air-carrier aircraft with humidity and icing sensors. These data will then be relayed automatically to the ground, in real time.) Now add the fact that the icing phenomenon is transitory in nature. From all this one must conclude that of all the aviation weather hazards, aircraft icing is one of the most difficult to forecast. From time to time, aviation publications malign what is perceived to be a shotgun approach to icing forecasts. Most often the criticism is unwarranted, usually based on a misunderstanding of forecast products and their purpose and scope. Almost never is any practical suggestion for improvement offered. This chapter examines icing forecasts and their purpose, scope, and interpretation. The final section deals with efforts to improve this valuable but often criticized product.

In today's aviation weather system, icing is directly forecast as AIRMET ZULU—how appropriate, since *Z* is often used to represent the freezing level—in the AIRMET Bulletin (WAs), SIGMETs (WSs), and Center Weather Advisories (CWAs). AIRMET ZULU provides forecasts for light to moderate icing, along with the forecasted freezing level. SIGMETs contain forecasts for severe icing. CWAs are issued for icing conditions not yet advertised in the AIRMET Bulletin or as a SIGMET or too localized to warrant an advisory.

Pilots must never overlook other forecast products that imply icing or provide additional or supplemental information on potential icing. These include Winds and Temperatures Aloft Forecasts (FDs), the Area Forecast (FA), and TWEB Route Forecasts. All these products, like the direct icing forecasts, are short-term products—less than 24 hours, except that the FD does go to about 30 hours.

For longer-term icing information, pilots have access to significant weather prognostic charts. These charts provide an outlook for potential icing out to about 48 hours.

As mentioned, the final section of this chapter contains a look into the future. Several new approaches to improve icing forecasts are discussed, both spatially (area of coverage) and temporally (time frames). These improvements are just around the corner. They should be making a significant operational impact around the year 2000.

Weather Advisories

During the latter part of the 1950s, the U.S. Weather Bureau issued warnings of potentially hazardous or severe aviation weather in the form of flash advisories. These were subsequently divided into AIRMETs (WAs) and SIGMETs (WSs). AIRMETs and SIGMETs alerted pilots that significant, previously unforecast weather had developed. Twenty-one U.S. Weather Bureau offices routinely issued advisories. By 1970, the number of flight advisory weather service offices was reduced to nine. Although "...the Inflight Weather Advisory program is intended to provide advance notice of potentially hazardous weather developments to en route aircraft...," according to the *Weather Service Operations Manual,* advisories were issued even when conditions were forecast in the Area Forecast. It was not uncommon to have four or more AIRMETs continuously in effect, and to this end, the Continuous AIRMET (WAC) was developed. Weather advisories issued by adjacent offices were not always consistent and often overlapped.

The Area Forecast was changed in 1978 to include a HAZARDs or flight precautions section to reduce the number of advisories. The FA was still issued by local NWS offices. Responsibility for issuing AIRMETs and SIGMETs for the 48 contiguous states was centralized in 1982 in Kansas City at the National Aviation Weather Advisory Unit (NAWAU).

In 1991, AIRMET criteria phenomena were removed from the Area Forecast and issued separately as the AIRMET Bulletin. This eliminated the redundancy of AIRMET/Flight Precautions and the disappearance of AIRMETs at the next FA issuance.

In October 1995, NAWAU became the Aviation Weather Center (AWC). Center Weather Service Units (CWSUs) were established at Air

Route Air Traffic Control Centers (ARTCC) in 1980. The purpose of the CWSU is to assist controllers and flow control personnel and alert pilots of hazardous weather through a Center Weather Advisory (CWA).

Conditions must be widespread for inclusion in an AIRMET Bulletin or a SIGMET; i.e., The condition must be occurring or forecast over an area of at least 3000 square miles, approximately three times the size of Rhode Island. Localized occurrences do not warrant the issuance of an AIRMET or SIGMET.

What's local? According to Dick Williams, AWC forecaster, AIRMET Bulletins and SIGMETs "...written on the scale of whole states do not endeavor to describe every single occurrence of IFR, icing, or turbulence. The forecaster, wishing to indicate that there may be isolated observations, pilots reports, or hazardous weather, may use the term 'local.' No hard and fast rule exists for determining when 'local' becomes widespread. The forecaster relies on observations, pilot reports, satellite imagery, and his or her own judgment in determining the extent of weather features." Federal Aviation Regulations recognize the fact that every occurrence of adverse weather cannot be forecast. Regulations require that even student pilots receive instruction in "The recognition of critical weather situations...." In other words, the excuse, "They didn't tell me," is just that, an excuse, not a reason.

Because of the widespread and transitory nature of icing, forecasters often use the conditional term *occasional* (*OCNL*) in the AIRMET Bulletin and SIGMETs. It is important for pilots to understand this term. *Occasional* is defined as a greater than 50 percent chance of occurrence but occurring for less than one-half of the forecast period.

Pilots must use all their training and experience when applying weather forecasts, especially those involving icing (more about this in Chapters 5 and 6).

Forecasts of icing begin with an analysis of current conditions. Upper-air data are analyzed for the lapse rate, temperature, and dew point. An empirically derived percentage frequency of icing for each layer is determined. Bases and tops with the highest probability of icing are plotted, along with the type of icing. Clear ice is indicated if the sounding is conditionally unstable; rime ice is indicated if the sounding is absolutely stable. Surface plots depicting clouds and weather, along with visual and infrared (IR) satellite imagery and radar, help the forecaster determine the horizontal and vertical extent of cloud and precipitation systems. PIREPs help the forecaster refine and modify the vertical and horizontal extent of the icing forecast.

As we shall see, the addition of computer models is helping forecasters to further improve weather advisory spatial coverage, altitudes, and severity.

TIP

Often errors or misunderstandings of icing forecasts are due to one or more of the following:

- The phenomenon is moving through the area of weather advisory coverage.
- The weather system is moving faster or slower than forecast.
- The forecaster uses the term occasional, because of the transitory nature of icing.
- The condition is localized and does not warrant an advisory.

AIRMET ZULU

AIRMET ZULU describes the location, intensity, and type (rime, clear, or mixed) of nonconvective icing. Convective (thunderstorm) activity always implies severe icing; a separate weather advisory will not be issued for convective icing. However, if the convective activity is embedded in stratiform clouds, a weather advisory will be issued. Layers where significant icing can be expected are expressed as specific values or ranges with bases and tops.

AIRMET ZULU also includes forecasts for light or local moderate icing. Light or local moderate icing will be described using geographic areas (*NE AZ AND NW NM*, etc.). Trace or no icing is indicated by the statement *NO SGFNT ICING EXPCD* or *NO SGFNT ICING EXPCD OUTSIDE CNVTV ACTVTY*. A separate paragraph contains forecast freezing levels. Terms such as *sloping* or *lowering* describe varying levels. The following is an example of this type of occurrence:

SLC WA 141345
AIRMET ZULU UPDT 2 FOR ICG AND FRZLVL VALID UNTIL 14200
NO SGFNT ICG XPCD OUTSIDE OF CNVTV ACTVTY.
OCNL LGT-ISOLD MOD RIME ICGIC BTN 150 AND FL220 OVER
SERN WY-WRN/CNTRL CO-WRN NM AND ERN AZ.

Here, the forecaster is warning of light to isolated moderate rime icing in clouds between 15,000 and 22,000 ft over southeastern Wyoming, western and central Colorado, and western New Mexico and eastern Arizona. Pilots of aircraft without ice protection equipment should note that trace icing is not forecast, even though trace icing could present a hazard to these aircraft.

Because of the extreme hazard presented by supercooled large droplets (SLDs), the Aviation Weather Center (AWC) has revised its icing advisories—AIRMETs and SIGMETs—to imply hazards due to SLD occurrences.

To alert pilots of potential SLD occurrences aloft, the AWC will issue advisories containing the terms *MIXED OR CLEAR ICING IN CLOUDS OR PRECIPITATION (MXD/CLR ICICIP)* or *CLEAR ICING IN PRECIPITATION (CLR ICGIP)*, which means precipitation-sized drops aloft. At times when the forecaster cannot separate rime and clear ice, the AIRMET will contain *RIME/MXD ICGICIP. FZDZ/RA ALF* (freezing drizzle and freezing rain aloft) also indicates the presence of SLDs. Even though the rate of accumulation is only moderate, the presence of SLDs poses a significant hazard, even to aircraft with ice protection equipment.

To indicate areas of SLDs, there may be an AIRMET within an AIRMET to highlight the threat. For example, an AIRMET for moderate rime icing below 14,000 ft may cover a relatively large area. A second AIRMET wholly within the first AIRMET's coverage may forecast moderate mixed or clear icing in clouds and precipitation below 10,000 ft. This alerts pilots to the SLD threat within the second area below an altitude of 10,000 ft, with a potential of cloud-size droplets between 10,000 and 14,000 ft.

Because most of today's aircraft, certified for flight into known icing, are capable of climbing through icing layers and flying well above potential icing areas, icing PIREPs are not as plentiful as one would like. PIREPs are so far the only means of validating the forecast.

DECISION-MAKING PROCESS

The following illustrates the decision-making process with forecast icing. The aircraft was a turbo Mooney on a flight from Bakersfield to Hayward, California. The synopsis indicated moisture but a stable air mass. Bases along the route were reported at around 5000 ft and tops at 9000 to 11,000 ft; the freezing level was at 7000 ft. Except for the coastal mountains, terrain along the route was close to sea level—terrain is a very important factor. The flight was planned at 12,000 ft, because the tops were relatively low and the aircraft had the performance to climb quickly through the potential icing layer. Just prior to departure, a PIREP was obtained from a landing air carrier confirming the tops. During the climb, trace to light icing was encountered, once on top, the ice sublimated quickly. On top, there were some buildups above 12,000 ft. Deviations to avoid these clouds were obtained from ATC. By circumnavigating the buildups, icing and turbulence were avoided. Should this be attempted in an airplane with lesser performance? Absolutely not! The bases and tops were known quantities. The airplane had the performance to climb quickly on top. Had this not been possible, the pilot had the option to return; cloud bases were over 4000 ft above terrain and well below the freezing level. Was the icing forecast correct? Yes. Remember the definition of *occasional*. Had emergency assistance been required, the pilot would have been a candidate for a violation. I do not intend to imply that this procedure is recommended. The decision rests solely with the pilot, based on the pilot's training and experience and the capability of the aircraft.

The following is another example of AIRMET ZULU:

SFOZ WA 141345
AIRMET ZULU UPDT 2 FOR ICG AND FRZLVL VALID UNTIL 142000
AIRMET ICG...WA OR CA

FROM YQL TO GGW TO BFF TO ALS TO 120W SFO TO 120W FOT TO 120W TOU TO YQL

OCNL MOT RIME/MXD ICGICIP BTN 040 TO 140 WA BTN AND 060 TO 160 OR/CA. CONDS CONTG BYD 20Z THRU 02Z.

FRZLVEL..WA W OF CASCDS..045 LWRG BY 20Z TO 035. WA CASCDS
   EWD..AT/NEAR SFC WITH MULT FRZLVLS 30-35.

OR W OF CASCDS..55 NORTH TO 65 SOUTH. AFT 18Z 50
   NORTH TO 75 SOUTH. OR CASCDS
   EWD..AT/NEAR SFC WITH MULT FRZLVLS
   TO 40-50
CA..AT/NEAR SFC SIERRAS AND NE PTN TIL 18Z. ELSW
   NEAR 70 NORTH SLPG TO 90 SOUTH. AFT 18Z 70
   NORTH SLPG TO 90 CNTRL AND 100 SOUTH.

The specific area extends from Lethbridge, Alberta (*YQL*), to Glasgow, Montana (*GGW*), to Alamosa, Colorado (*ALS*), to 120 nautical miles west of San Francisco, California (*SFO*), to 120 nautical miles west of Fortuna, California (*FOT*), to 120 nautical miles west of Tatoosh, Washington (*TOU*), and back to *YQL*. This area includes the coastal waters.

Intensities, type, and altitudes may differ significantly over the areas affected. Occasional moderate rime or mixed icing in clouds and precipitation is forecast from 4000 to 14,000 ft MSL over Washington and from 6000 to 16,000 ft MSL over Oregon and California. This forecast implies SLD potential in these locations. Conditions are expected to continue beyond the end of the forecast period (20Z), through the outlook period (0200Z). Pilots planning flights beyond the forecast period of 20Z can expect this AIRMET to be in effect at least through 02Z.

It seems a bit redundant to include the base of the icing in the icing paragraph and freezing level in the subsequent paragraph. This resulted from a pilot obtaining a briefing with icing from the *FRZLVL*. The base of the freezing level was not specified. Now the forecaster must enter a specific altitude in the icing paragraph. This may be in the form of *BTN,* as in the preceding example, or *FRZLVL-150. FRZLVL 55-70.* Pilots can expect the most significant icing within the layer specified in the icing paragraph. Trace or light icing can be expected between the freezing level and the altitude in the icing paragraph.

Next appears the freezing level paragraph (*FRZLVL*). In Washington, west of the Cascades, the freezing level is expected to lower from 4500 to 3500 ft MSL by 2000Z. East of the Cascades, the freezing level is at or near the surface, with multiple freezing levels between the surface and 3000 to 3500 ft MSL. Multiple freezing levels are caused by overrunning warm air, such as with a warm front or warmer air overrunning colder air trapped in valleys. Freezing precipitation occurs in these areas, a definite indicator of SLDs. Similar conditions are expected in Oregon but at different levels and times. One would expect the greatest potential for SLDs in Washington and Oregon east of the Cascade Mountains.

In California, the freezing level is forecast at or near the surface in the Sierra Nevada Mountains and northeast portion until 1800Z. Elsewhere, the freezing level is expected near 7000 ft MSL in the north sloping to 9000 ft MSL in the south. After 1800Z, the freezing level is forecast to remain around 7000 ft in the north and rise in the central and southern portions to 9000 to 10,000 ft.

In this section the forecaster has described an icing layer 6000 to 10,000 ft deep that slopes upward about 2000 ft from north to south. The type of ice forecast, mixed, and the depth of the anticipated icing layer indicate an unstable air mass. In the standard atmosphere this represents a temperature drop of approximately 12 to 20°C. From the discussions in Chapter 1 this is typically the most hazardous icing temperature range.

IFR pilots with aircraft certified for flight in icing conditions should have little trouble in these areas, assuming performance will allow them to climb out of the icing layer, although they must be prepared to contend with SLDs and freezing rain east of the Cascades and icing to the surface in parts of California. For IFR pilots of aircraft without ice protection equipment, this forecast would be a very strong *no-go* indicator, especially east of the Cascades and the mountains of California. The VFR pilot flying in Washington or Oregon east of the Cascades will be just as susceptible to icing as an IFR pilot because of multiple freezing levels and possible freezing rain. However, any flight decision cannot be based on one forecast, especially taken out of context, as in this case.

When a SIGMET has been issued, the AIRMRT Bulletin alerts pilots as illustrated in the following example:

```
SFOZ WA 031740 AMD
AIRMET ZULU UPDT 4 FOR ICE AND FRZLVL VALID UNTIL 032100
...SEE SIGMET NOVEMBER SERIES FOR POSS SEV ICE...UPDT...
AIRMET ICE...CA AND CSTL WTRS NV AZ
FROM LKV TO EKO TO YUM TO 210WSW SAN TO 120W FOT TO LKV
OCNL MOD ISOL SEV RIME/MXD ICGICIP BTN 070 AND FL200.
```

In this example, the forecaster has alerted the pilot to the potential of isolated severe rime or mixed icing. Even though severe icing is forecast, coverage is only expected to be isolated—local. Therefore, this does not qualify as a SIGMET. However, the AIRMET Bulletin was amended (*AMD*) at 1740Z to refer users to *SIGMET NOVEMBER SERIES FOR POSS SEV ICE*. The severe icing is no longer localized.

Icing SIGMETs

SIGMETs, like the AIRMET Bulletin, often cover large areas because of the widely scattered and transitory nature of the phenomena they report. Therefore, the term *occasional* (*OCNL*) appears frequently. Additionally, phenomena may move through or only affect certain geographic features within the advisory area. Failure to completely read and understand an advisory has lead many a pilot and FSS briefer to unjustly criticize the product. Again, like the AIRMET Bulletin, SIGMETs are only issued when a phenomenon is widespread. Local occurrences of severe icing appear in AIRMET Bulletin.

In Alaska and Hawaii, weather service forecast offices responsible for Area Forecasts (FAs) issue SIGMETs. SIGMETs are identified by forecast area, alphabetic, and product designators. The forecast area specifies within which FA the advisory applies. Next appears the alphabetic designator for the phenomenon being described. (To avoid confusion with International SIGMETs, domestic SIGMET names now run *NOVEMBER* through *YANKEE*—excluding *SIERRA, TANGO,* and *ZULU,* which are reserved for AIRMETs.) The product designator (1, 2, 3, etc.) indicates the number of successive times the advisory has been issued. To ensure continuity and alert pilots and controllers, a referencing remark may be appended to the message (*FOR PREVIOUS ISSUANCE SEE SLC OSCAR 2*). Updates often contain changes; they must be reviewed for affected area, altitudes, and times. It is important to note both phenomenon and product designators.

```
SFON UWS 031715
SIGMET NOVEMBER I VALID UNTIL 032115
WA OR
FROM 20W BLI TO 30E BLI TO 20SE PDX TO 40WSW PDX TO 20W BLI
MOD OCNL SEV RIME/MXD ICGICIP BTN 090 AND 140. RPTD BY
SVRL ACFT.
CONDS CONTG BYD 2115Z.
```

```
SFOO UWS 031815
SIGMET OSCAR I VALID UNTIL 032215
WA OR
FROM 20W BLI TO 30E BLI TO 20SE PDX TO 40WSW PDX TO 20W BLI
MOD OCNL SEV RIME/MXD ICGICIP BTN 090 AND 140. RPTD BY
SVRL ACFT.
CONDS CONTG BYD 2215Z.
...THIS REPLACES SIGMET NOVEMBER FOR THE SFO FA AREA...
```

The first example is San Francisco SIGMET NOVEMBER 1. Like the AIRMET Bulletin, the following lines provide the states and locations within the states where the phenomenon is expected. Moderate to occasional severe rime or mixed icing in clouds and precipitation is forecast between 9000 and 14,000 ft. The SIGMET is in part based on PIREPs (*RPTD BY SVRL ACFT*). This illustrates the importance of accurate pilot weather reports.

The second example is San Francisco SIGMET OSCAR 1. Location and conditions are the same, and the forecaster has noted that this replaces SIGMET NOVEMBER for the San Francisco FA area. (Most likely SIGMET NOVEMBER was already being used, and the forecaster had to change the name to prevent any misunderstanding. The forecaster has, however, alerted us to the change.) Note that in both advisories the forecaster alerts pilots to the potential for SLDs (*/MXD ICGICIP*).

Icing Center Weather Advisories

Center Weather Advisories (CWAs), unscheduled inflight advisories, are issued when conditions are expected to significantly affect IFR operations to help pilots avoid hazardous weather. The advisories update or expand the AIRMET Bulletin, SIGMETs, or the Area Forecast and may be issued when conditions meet advisory criteria. In such cases, the CWSU will coordinate with AWC forecasters for the issuance of the appropriate advisory. CWAs are also issued when local hazardous conditions develop that do not warrant other advisories.

The CWA numbering system was somewhat complex but has been simplified. CWAs have a three-digit number. The first digit is a phenomenon number, i.e., a specific weather event that required issuance of the CWA. A separate phenomenon number is assigned each distinct condition (turbulence, icing, thunderstorms, etc.). For example, *101* may forecast a turbulence event; *201* may be issued for icing. The second and third digits indicate the number of times a specific phenomenon has been updated, e.g., *101* (first issuance), *102* (second issuance), and so on.

The following is an example of an icing CWA:

ZKC CWA 102 VALID UNTIL 161645
FROM PWE TO BUM TO EVV TO 30E DEC TO BRL TO PWE

AREA OCNL MOD RIME ICE RPTD BTN FL180 AND FL240
MAINLY NEAR KMKC. CONDS CONT BYD 161645.

This Kansas City Center CWA for the northern third of Missouri and central Illinois describes an area of occasional moderate rime icing from 18,000 to 24,000 ft. The hazard has been reported by aircraft. The forecaster indicates that the phenomenon is mainly near Kansas City and will continue beyond 161645Z.

This area of icing was not carried in the normal update of the AIRMET Bulletin and was considered significant enough for the CWSU forecaster to issue a CWA. As is often the case, the AIRMET Bulletin was then amended and the CWA subsequently canceled or not reissued. Typically, when this occurs, the CWSU will issue a cancellation, referring users to the appropriate updated AIRMET Bulletin or SIGMET.

Dissemination

Advisories are provided routinely during FSS standard briefings and are offered during abbreviated briefings. (FSS weather briefings will be further discussed in Chapter 5.)

During routine FSS radio contacts, advisories within 150 mi will be offered when they affect the pilot's route. It is important to note SIGMET series and number to ensure receipt of the latest information.

In the contiguous United States, the Hazardous Inflight Weather Advisory Service (HIWAS) has been commissioned to broadcast weather advisories and urgent PIREPs continuously over selected VORs. HIWAS facilities are advertised on aeronautical charts and in the *Airport/Facility Directory*.

When an advisory affects an area within 150 mi of an HIWAS outlet or an ARTCC sector's jurisdiction, an alert is broadcast once on all frequencies—except Flight Watch and emergency. Approach controls and towers also broadcast an alert, but it may be limited to phenomena within 50 mi of the terminal. When the advisory affects operations within the terminal area, an alert message will be placed on the ATIS. Overzealous tower controllers have been known to place SIGMET alerts for conditions hundreds of miles away on the ATIS.

Despite criticism that advisories cover too much area, their issuance has become more conservative. Ironically, some pilots and FSS controllers now criticize the forecast for not containing enough precautions. Virtually all criticism, however, is due to misconceptions and misunderstanding the product.

The existence of an advisory, or lack thereof, does not relieve the pilot from using good judgment and applying personal limitations. Like all pilots, I have on occasion had to park my turbo Cessna 150 and take one of American's Boeing 727s. These instances lend credence to the aviation axiom "When you have time to spare, go by air; more time yet, take a jet." When you do not have the equipment or qualifications to handle the weather, *do not go!* This does not mean that every time you hear an advisory, you cancel, but you must take a close look at all available information.

CASE STUDY

For a flight from Las Vegas, Nevada, to Van Nuys, California, I was told by the briefer, "Well, you aren't going today!" My jaws locked up, and I replied, "Oh yes I am!" I hadn't looked at the weather yet; my statement was a gut reaction to this individual's horrible attitude. Advisories for turbulence, icing, and mountain obscurement were in effect, along with a forecast for marginal VFR and rain showers. As is often the case in this part of the country, a direct flight was out. However, by choosing a course over lower terrain, VFR is frequently possible. My decision was based on my experience, knowledge of the terrain, a thorough review of all available weather reports and forecasts, and always having an out should the weather ahead become impassable.

> **CASE STUDY**
> The lack of an advisory does not guarantee the absence of
> hazardous weather. An unfortunate pilot learned this lesson the
> hard way. The synopsis described a moist unstable air mass.
> Thunderstorms were not forecast for the time of flight but were
> expected to develop; thunderstorms, however, were already
> being reported along the route. The pilot, without storm-
> detection equipment, encountered extreme turbulence
> inadvertently entering a cell. The pilot, with three passengers,
> filed an IFR flight plan based on the fact that there were no
> advisories. The pilot had the clues—moist unstable air,
> thunderstorms already reported—but put complete trust in a
> forecast that included no advisories.

The case studies illustrate two go decisions. One resulted in a
routine flight; the other was almost fatal. My intent is not to brag
about my skills or criticize another individual. I hope to show the
process, based on available information and a knowledge of weather
products and limitations, that led to the decisions.

I have already touched on the point that if you wish to be accorded
and exercise the privileges of pilot-in-command, you must understand
the system and its limitations. You must evaluate all available
information—as required by regulations—and make a flight decision
based on your knowledge and limitations and that of your aircraft and
its equipment.

Terminal Aerodrome Forecasts (TAFs)

TAFs are issued four times a day (0000Z, 0600Z, 1200Z, and
1800Z) and are valid for 24 hours. Forecasts using TAF (World
Meteorological Organization, WMO) codes are also issued for
many military locations by local base weather offices. Military
TAFs have different criteria from those issued by the National
Weather Service (NWS); therefore, some differences will
appear.

> Aviation in itself is not
> inherently dangerous. But to an
> even greater degree than the
> sea, it is terribly unforgiving of
> any carelessness, incapacity, or
> neglect.

TAFs are issued in the following format:

- Type
- Location
- Issuance time
- Valid time
- Forecast

There are two types of TAFs, a routine forecast, *TAF*, and an amended forecast, *TAF AMD*. TAFs may be corrected (*COR*) or routinely delayed (*NIL*). TAF location is identified by the four-letter ICAO station identifier. Issuance date and time consist of a six-digit group. The first two digits represents the day of the month, and the last four digits represent UTC issuance time. The valid period is a four-digit group in UTC, usually for a period of 24 hours. The body of the TAF uses the following format:

- Wind
- Visibility
- Weather
- Sky condition
- Wind shear, when applicable

Like the METAR code, wind is forecast as a five- or six-digit group when considered significant to aviation. The contraction *KT* follows the wind forecast and denotes the units as knots. Gusts are noted by the letter *G*.

Pilots need to be aware of strong surface winds. Winds may limit or preclude the use of certain runways, especially those with a significant crosswind. If snow, ice, or slush covers available runways, length may be another significant factor. If fresh snow has fallen, a forecast of strong surface winds indicates the possible blowing or drifting of snow.

Prevailing visibility up to and including 6 statute miles is forecast. Visibility greater than 6 miles is indicated by the letter *P* for plus (*P6SM* means visibility greater than 6 statute miles). Military and many international TAFs forecast visibility in meters.

Low ceilings, along with low visibilities, can increase landing hazards with ice on the aircraft or the surface. Poor visibility with a ceiling and snow falling can produce a whiteout effect. Both VFR and IFR pilots should consider increasing minimums above "legal" levels during these conditions.

Weather and obstructions to vision use the same format and codes as METAR (see Table 3-1). With no significant weather expected, the weather group is omitted. When significant weather is forecast and then expected to change in the future to no significant weather, the contraction *NSW* (no significant weather) appears. *NSW* will not appear when any of the following phenomena are expected to occur:

- Freezing precipitation

- Moderate or heavy precipitation

- Drifting or blowing dust, sand, or snow

- Duststorms or sandstorms

- Thunderstorms

- Squall

- Tornadoes

- Phenomena expected to cause a significant change in visibility

Look for significant icing indicators, such as freezing or frozen precipitation. Do not overlook the possibility of serious icing with liquid precipitation falling at the surface. Recall from previous chapters how liquid surface precipitation, under certain conditions, is a positive indicator of icing aloft. Forecasts of blowing or drifting snow may restrict or preclude the use of an airport. A decision to make a flight should be predicated on an alternate airport not under the influence of the weather system affecting the destination airport. That is, you may have a "legal" alternate at an airport 10 mi away. However, if the forecast goes "belly up" at the destination, most likely the same conditions will affect the alternate.

Sky condition uses the same format and contractions as METAR, i.e., amount, height, cloud type, or vertical visibility. When cumulonimbus clouds are expected, *CB* is appended to the cloud layer. *CB* is the only cloud type forecast in TAFs.

Nonconvective low-level wind shear, when forecast, will appear following sky condition on domestic TAFs, in the following format: *WS015/24035KT.*

- *WS* means wind shear.

- *015* is the height in hundreds of feet (above ground level) of the wind shear (1500 ft AGL).

- */24035* is wind direction and speed (knots) above the wind shear (240° at 35 knots).

A forecast of low-level wind shear may be another no-go indicator. Carefully consider the advisability of making a flight to a snow-, ice-, slush-covered airport with the additional factor of a wind shear encounter. Negative factors must be considered cumulative. While one may be considered within a safe margin, each additional factor exponentially increases risk. I will talk more about this in Chapter 5 in discussions of risk assessment and management.

Conditional Terms

TAF conditional terms consist of temporary (*TEMPO*) conditions and probability (*PROB*) forecasts. *TEMPO* indicates that temporary conditions are expected to occur during the forecast period. *TEMPO* describes any condition generally expected to last for less than an hour at a time. The time during which the condition is expected to occur is indicated with a four-digit group giving beginning and ending time UTC. For example, *SCT030 TEMPO 1923 BKN030* means at 3000 ft, scattered temporarily between 1900Z and 2300Z, ceilings at 3000 ft and broken. A ceiling at 3000 ft and broken is expected to exist for periods of less than 1 hour during the 1900Z to 2300Z time frame. *TEMPO* is equivalent to and often translated by FSS controllers as *occasional* (*OCNL*).

A *PROB* group indicates the probability of occurrence of thunderstorms or other precipitation events. This is followed by a four-digit time group giving beginning and ending times. *PROB40* is the equivalent to *CHC,* and *PROB30* is equivalent to *SLT CHC.* They represent a 40 or 30 percent probability of occurrence, respectively.

Forecast Change Groups

Forecast change groups consist of from (*FM*) followed by a time group (*tttt*) and becoming (*BECMG*) followed by a time group (*TTtt*). The *FMtttt* group is used when a rapid change is expected, usually less than 1 hour. For example, *BKN020 FM1630 SKC* means before 1630Z, ceiling at 2000 ft and broken; around 1630Z, the sky condition will change to clear. The *BECMG TTtt* group indicates a more gradual change in conditions over a longer period of time, usually 2 hours. For example, *5SM HZ BKN 030 BECMG 0507 3SM BR OVC 020* means before 0500Z, visibility is 5 statute miles in haze, ceiling at 3000 ft and broken; then during the period of 0500Z to 0700Z, conditions changing to visibility 3 statute miles in fog, ceiling at 2000 ft and overcast.

Below is an example of a TAF for Seattle Tacoma International Airport:

```
TAF
KSEA 041045Z 1212 00000KT P6SM SCT015 OVC025 TEMPO 1214 -RA
     BECMG 1314 24007KT SCT015 BKN025 TEMPO 1418 -SHRA
     BECMG 1718 SCT020 BKN035 PROB40 1801 -TSRA BECMG 0001
     22006KT SCT015 BKN035 TEMPO 0112 -SHRA=
```

This TAF was issued for Seattle Tacoma (*KSEA*) on the fourth day of the month at 1045Z (*041045Z*). The forecast is valid from the fourth at

1200Z until the fifth at 1200Z (*1212*). At 12Z, the surface winds are expected to be calm (*00000KT*). Visibility forecast is more than 6 statute miles (*P6SM*). Sky condition at 1500 ft is scattered, ceiling 2500 ft and overcast. Occasionally (*TEMPO*) between 1200Z and 1400Z (*1214*), light rain is expected. Prevailing conditions are expected to become (*BECMG*) between 1300Z and 1400Z (*1314*) wind 240 at 7 knots (visibility is implied to be more than 6 miles), 1500 ft scattered, ceiling 2500 ft and broken. Occasionally, between 1400Z and 1800Z, light rain showers are expected. Between 1700Z and 1800Z (wind and visibility are implied to be 240 at 7 knots and more than 6 miles), sky condition at 2000 ft is scattered, with ceiling at 3500 ft and broken. There is a chance (*PROB40*) of thunderstorms with light rain. Between 0000Z and 0100Z, wind 220 at 6 knots, visibility more than 6 miles, sky condition at 1500 ft is scattered, and ceiling at 3500 ft and broken; occasionally, between 0100Z and 1200Z, light rain showers are forecast.

Below is an example of the Tacoma, Washington, McCord Air Force Base forecast for the same time period as the preceding example:

```
TCM 1212 20009KT 9999 SCT010 BKN020 OVC050 T12/13 540109
   610703 QNH2980INS CIG020
   BECMG 1314 20009KT 9999 SCT015 BKN035 OVC045 540109
   620703 QNH2975INS CIG035
   BECMG 1920 24007KT 9999 SCT020 BKN045 BKN150 T20/23
   540009 620505 QNH2980INS CIG045
   BECMG 0001 VRB05KT 9999 -SHRA SCT020 BKN040 OVC070
   610505 QNH2983INS CIG040
   BECMG 0405 VRB05KT 9999 NSW SCT020 BKN050
   QNH2990INS CIG050=
```

There are several major differences between domestic TAFs and those issued by the military and foreign governments. Visibilities forecasts are in meters. In the preceding example, the visibility through the forecast period is *9999*—greater than 9000 m. Sky condition forecasts do not use the summation principle, and the ceiling is specified (*BKN015 OVC080 BKN200 CIG015*). The altimeter setting is forecast (*QNH2980INS* means altimeter setting at 29.80 inches of mercury).

Maximum and minimum temperatures, icing, and turbulence are sometimes forecast. Refer to Table 4-1. The temperature group follows sky conditions. For example, *T12/13* means minimum temperature of 12°C at 1300Z; *T20/23* means maximum temperature of 20°C at 2300Z. In the example, at 1200Z, *540109* appears. The 5 (*5*40109) represents the turbulence group. The 4 (5*4*0109) indicates the turbulence intensity—moderate turbulence in cloud, infrequent. The 010 (54*010*9) shows the base of the turbulence layer height in hundreds of feet—1000 ft. The 9 (540109) represents the thickness of the turbulence layer in thousands of feet—9000 ft. Moderate turbulence is forecast between 1000 and 10,000 ft. In the example, at 1200Z, *610703* appears. The 6 (*6*10703) represents the icing group. The 1 (6*1*0703) indicates the icing intensity—light icing. The 070 (61*070*3) shows the base of the icing layer height in hundreds of feet—7000 ft MSL. The 3 (61070*3*) represents the thickness of the icing layer in thousands of feet—3000 ft. Light icing is forecast between 7000 and 10,000 ft.

Notice in Table 4-1 icing intensities 6 and 9. Both these categories would indicate potential SLDs. Since turbulence and icing forecasts are provided in the AIRMET Bulletin and SIGMETs, U.S. domestic TAFs do not contain these phenomena.

Winds and Temperatures Aloft Forecasts

Winds and Temperatures Aloft Forecasts (FDs) for the contiguous United States, Alaska, and many oceanic areas are computer generated at the National Meteorological Center outside Washington, D.C. FDs for the Hawaiian Islands are produced by the Honolulu forecast office. These forecasts are based on twice-daily radiosonde balloon observations, normally released at 1100Z and 2300Z daily.

Table 4-1.
Terminal Aerodrome Forecasts (TAF)

Temperature: TT_IT_I/tt

T	-Temperature Group
T_IT_I	-Temperature in Celsius
tt	-Time UTC

Turbulence: 5ihhhd

5	-Turbulence Group
i	-Turbulence Intensity
hhh	-Base Height hundreds of feet
d	-Thickness in thousands of feet

Icing: 6ihhhd

6	-Icing Group
i	-Icing Intensity
hhh	-Base Height hundreds of feet
d	-Thickness in thousands of feet (0 indicates to top of clouds)

Turbulence Intensity
0 None
1 Light turbulence
2 Moderate turbulence in clear air, infrequent
3 Moderate turbulence in clear air, frequent
4 Moderate turbulence in cloud, infrequent
5 Moderate turbulence in cloud, frequent
6 Severe turbulence in clear air, infrequent
7 Severe turbulence in clear air, frequent
8 Severe turbulence in cloud, infrequent
9 Severe turbulence in cloud, frequent

Icing Intensity
0 No icing
1 Light icing
2 Light icing in cloud
3 Light icing in precipitation
4 Moderate icing
5 Moderate icing in cloud
6 Moderate icing in precipitation
7 Severe icing
8 Severe icing in cloud
9 Severe icing in precipitation

Approximately 750 upper-air stations worldwide—about 120 in the United States—make observations. Stations are generally located on land, leaving great expanses of ocean without observations; satellite and aircraft reports help fill in the gaps. The computer must interpolate for locations without observations, and sparse observational data hinder the accuracy of the forecast.

FDs normally become available after their scheduled transmission times of 0440Z and 1640Z. They consist of three forecast periods: 6, 12, and 24 hours. These periods are labeled *FD1*, *FD2*, and *FD3* for levels through 39,000 ft and *FD8*, *FD9*, and *FD10* for 45,000 and 53,000 ft.

The following is an example of a tabulated Winds and Temperatures Aloft Forecast through 30,000 ft.

```
DATA BASED ON 250000Z
VALID 260000Z    FOR USE 1800-0500Z.      TEMPS NEG ABV 24000
FT   3000    6000     9000    12000   18000   24000   30000
SAC 2122 2229+07 2337+04 2444+00 2459-13 2357-23 236039
RBL  2030 2141+04 2349+02 2457-02 2474-15 2378-24 238639
SIY       2156+03 2362+00 2467-04 2484-16 2495-25 740740
PDX 2431 2451+01 2363-04 2273-09 2398-20 7321-31 734643
SEA 2527  2441-03 2357-08 2278-12 7209-22 7331-33 735745
```

This forecast is based on the twenty-fifth day of the month 0000Z radiosonde data (*DATA BASED ON 250000Z*). The *DATA BASED ON* always must be checked. From time to time, old FDs fail to be purged and remain in the system. It is possible to receive data that is 24 hours old. The next line states *VALID 260000Z FOR USE 1800-0500Z*. This FD is for use between 1800Z and 0500Z. The computer does not forecast an average; the model predicts winds and temperatures for one specific time, in this case 0000Z (*VALID 260000Z*).

Forecasts based on expected movement of synoptic systems explain one reason for apparent errors. With rapidly moving or intensifying systems, FDs can change significantly during the *FOR USE* period.

Levels within the area of frictional effect between the wind and the earth's surface are omitted. Therefore, forecast levels within approximately 1500 ft of the surface and temperatures for the 3000-ft level or levels within 2500 ft of the surface do not appear. Pilots must interpolate—compute intermediate values—to determine values between forecast levels and reporting locations.

Note the Seattle 30,000-ft level. Is this forecast of *735745* a misprint or garbled transmission? With forecast winds of 100 knots or more, 5 is added to the first digit of the wind-direction group. Therefore, to decode, subtract 5 from the first digit of the wind direction, and add 100 to speed. In this example, wind direction, speed, and temperature are from 230° true, at 157 knots, temperature −45°C. [No mathematical sign (+ or −) was specified, so the temperature must be negative.] Maximum speed for FD tabulated forecasts is 199 knots.

Forecast Winds and Temperatures Aloft Forecasts are also available in graphic form, issued twice daily at 1200Z and 0000Z. FD charts are excellent for determining forecast winds for long-distance flights. Because winds at various levels are depicted visually, favorable routes and altitudes can be determined. The eight panels contain forecast levels from 6000 through 39,000 ft. Plotted data are standard.

Although FDs are generated in Washington, regional NWS offices are responsible for amendments. FDs are amended when, in the forecaster's judgment, there is a change or an expected change in the wind or temperature that would significantly affect aircraft operations. Amendment procedures are complex.

FDs provide the pilot with two valuable pieces of information: wind direction and speed plus temperature. Both significantly affect aircraft operation and performance. Failure to properly consider and apply either can be potentially hazardous.

Despite its limitations, the FD can never be ignored. Pilots are required by FARs to consider "...fuel requirements..." and are prohibited from beginning a flight either VFR or IFR "...unless (considering wind and forecast weather conditions)..." the aircraft will have enough fuel to fly to the destination or an alternate, if required, and still have appropriate fuel reserves. FAR fuel reserve minimums, which do not necessarily equate to "safe," in no way relieve the pilot from keeping careful track of ground speed and revising the flight plan accordingly.

FDs are a source of forecast freezing level. Refer the previous Winds and Temperatures Aloft Forecast example. Based on the standard atmosphere lapse rate of 2° per 1000 ft, the freezing level over Sacramento, California, is approximately at 12,000 ft. It lowers in northern California (*SIY*) to 9000 ft; in the Portland area, it is 5500 ft. By Seattle, the freezing level drops to about 4500 ft.

From a knowledge of potential icing, one can see that the top of the most significant icing should be between 16,500 and 19,000 ft in southern areas and 10,000 and 14,000 ft in northern areas. This would be the height of the −10 to −15 C temperatures. Additionally, one would expect no significant icing above about 30,000 ft in the south and 28,000 ft in the north—the height of −40°C temperatures. Keep in mind, however, that we have not taken into consideration other icing factors, such as cloud tops, stability, or wind shear.

These levels in general should be in agreement with the AIRMET Bulletin because they are based on the same data. Differences result from FD freezing levels representing only one point in time, whereas AIRMET Bulletin forecasts take into account changes during the period. Additionally, lapse rates may not be standard. If a significant difference occurs, be alert for other possible forecast errors. FSS controllers coordinate with forecasters under such circumstances. Pilots using DUATs normally would have to consult an FSS for resolution.

The situation may change in the future with wind profilers that automatically and almost continuously provide high-resolution upper-air wind, humidity, and temperature measurements. Naturally, the closer to observation time, the more accurate is the forecast. Accuracy normally deteriorates with time—the FD2s and especially the FD3s, although the FD3s are a source of extended forecast temperatures aloft and freezing level. Since the evening forecast does not become available until after 0440Z, it does not do much good to request winds and temperatures any earlier for the following day. Flights departing after 1640Z must consult the new FDs. Weather patterns can change significantly in 12 hours.

Other Forecast Products

Like TAFs, there are a number of additional weather products that imply icing. This section reviews the Area Forecast (FA), the TWEB Route Forecast (TWEB), and low-altitude significant weather prognostic charts—often simply referred to as *prog charts*.

As with surface reports and TAFs, we are looking for surface winds, visibilities, weather phenomena (liquid, freezing, frozen, and convective activity), and sky conditions that imply icing or increase the hazards of landing with ice or on a snow-, ice-, or slush-covered runway. An important new element available in the FA, and sometime in TWEBs, is forecast cloud tops. Unfortunately, as with all forecast products, there are limitations.

The Area Forecast

The Area Forecast (FA) predicts conditions over an area the size of several states. Because of limitations on size, computer storage, and communications equipment, the forecast cannot be divided into smaller segments nor provide the detail available in TWEBs or TAFs. Widely varying conditions over relatively large areas must be included; therefore, small-scale events are often described using conditional terms. The FA provides a forecast for the en route portion of a flight and destination weather for locations without TAFs. Conditions are forecast from the surface to 70 mb (approximately 63,000 ft). The FA provides a synopsis along with a VFR clouds and weather section.

The VFR clouds and weather section includes sky condition, non-IFR cloud heights and visibilities, obstructions to visibility, weather, surface winds, and a 6-hour categorical outlook.

Sky condition provides cloud height, amount, and tops. Heights are normally MSL, with AGL or ceiling (CIG) generally limited to layers within 4000 ft of the surface. Since tops of building cumulus, towering cumulus, and cumulonimbus clouds are quite variable, only upper limits appear (*TOPS FL350, CB TOPS FL300*, etc.). When multiple or merging layers are forecast that would not permit VFR flight between layers, only the top of the highest layer appears (*BKN-OVC80-100 LYRD TOPS FL200, MEGG/NMRS LYRS TOPS FL180*, etc.). Because of the scope of FAs, tops cannot be more precise. TWEB Route Forecasts may contain more detail.

Usually, combined with weather, surface visibility, and obstructions to vision are forecast when expected to be 5 miles or less. For example, *3SM-5SM -RA BR* means visibility 3 to 5 miles in light rain and mist, and *VSBY 3SM-5SM -SHRA PROB30 -TSRA* means visibility 3 to 5 miles in light rain showers and widely scattered light rain showers and thunderstorms. The absence of a visibility forecast only implies general visibilities greater than 5 miles. Since they are not within the scope of this product, widespread visibilities of 6 miles or local conditions at less than 5 miles may exist and not be included in the FA. TWEBs and TAFs often contain greater detail.

Widespread areas of strong surface winds considered operationally significant appear in the forecast (*20G30KT*, etc.). Direction is true, referenced to the eight points of the compass (*N, NE, E*, etc.). The lack of a wind forecast only implies widespread sustained speeds of less than 30 knots. TWEB Route Forecasts and TAFs again often can be used to determine winds of lesser speeds and local conditions. Often associated with convective activity, *TSRA G40KT* translates to wind gusts to 40 knots in the vicinity of thunderstorms and moderate rain showers.

A 6-hour categorical outlook (*OTLK*) appears at the end of each 12-hour VFR CLDS/WX statement. The *OTLK* consists of the following categories: IFR (Instrument Flight Rules), MVFR (Marginal Visual Flight Rules), and VFR (Visual Flight Rules); these categories do not necessarily correspond to FAR definitions but rather to those used on the weather depiction chart. For example:

- *IFR CIG* means ceiling less than 1000 ft.

- *IFR CIG BR* means ceiling less than 1000 ft and visibility less than 3 miles in mist.

- *MVFR HZ FU* means visibilities between 3 and 5 miles in haze and smoke.

- *VFR WIND* means ceiling greater than 3000 ft and visibility greater than 5 miles, sustained wind 30 knots or greater.

As well as from FSSs, the Area Forecast can be obtained through DUATs and other commercial vendors and on the AWC's Internet home page (*www.awc-kc.noaa.gov*).

TWEB Route Forecasts

TWEB aviation forecasts were developed as scripts for the FAA's Transcribed Weather Broadcast (TWEB) and Pilots Automatic Telephone Weather Answering Service (PATWAS). TWEBs also may be used on the Telephone Information Briefing Service (TIBS). TWEBs are issued three times a day and are valid for 15 hours.

In addition to Route Forecasts, NWS offices responsible for TWEBs prepare a synopsis. The synopsis contains a brief description of fronts, pressure systems, and local climate or terrain factors affecting the routes, valid for the same period as the route forecasts. The TWEB synopsis often contains more detail than the FA; therefore, it might be of more value in describing local conditions. More detail is possible because the TWEB synopsis only covers about one-fifth the area of the FA.

TWEBs use the same terminology and contractions as the FA. Forecasts contain significant clouds and weather, with *significant* defined as phenomena affecting at least 10 percent of a route or in the forecaster's judgment important to flight planning. Forecasts consist of strong surface winds, visibility, weather, including mountain obscurement, nonconvective low-level wind shear, and sky condition. Cloud types may be included.

Cloud heights are based on a standard reference: *ALL HGTS MSL XCP CIGS* (all heights mean sea level except ceilings) or *ALL HGTS AGL XCP TOPS* (all heights above ground level except tops). The use of AGL and CIGS is normally limited to layers within 4000 ft of the surface. TWEBs used to be basically a low-altitude forecast—below 15,000 to 18,000 ft. This no longer necessarily applies, but forecasters emphasize cloud conditions below 12,000 ft. With no clouds or scattered clouds above approximately 12,000 ft, the forecaster may use the phrase *NO SGFNT CLDS/WX* (no significant clouds or weather); this can be interpreted as any clouds present should be easily circumnavigable.

When forecast, tops normally will be included only for layers with bases below 12,000 ft. Like the FA, for multiple or merging layers that would not permit VFR flight between layers, only the top of the highest layer appears. However, because of the local nature of the product, tops are often more specific and, therefore, useful. Cloud tops also may be specified indirectly. For example, *CIG BLO 10 OBSCG TRRN BLO 15* means that cloud tops are expected to be 1500 ft.

Surface visibilities are forecast when expected to be 6 miles or less. However, to prevent misunderstanding, *P6SM* may be used to indicate unrestricted visibility. Weather and obstructions to vision appear as necessary.

Sustained surface winds normally are forecast when expected to be 25 knots or greater. Notice that the TWEB threshold for winds is 5 knots less than the FA. Direction is referenced to true north and given as cardinal headings. The lack of a wind forecast only implies sustained speeds of less than 25 knots over 90 percent of the forecast area.

Local Vicinity Forecasts have been developed to cover metropolitan areas. They normally cover a radius of 50 nautical miles.

TWEBs do not directly forecast turbulence or icing. However, turbulence and strong up- and downdrafts are implied by weather associated with these phenomena. For example, strong winds and mountain wave activity indicate turbulence. Icing can be expected above the freezing level where visible moisture exists or in areas of freezing precipitation. Like the FA, thunderstorms imply severe or greater turbulence and icing and low-level wind shear.

TWEB routes and synopses often provide more precise timing and detail than are possible in the FA. They provide additional specific information on visibility, surface winds, and local conditions not possible in the FA. However, like other forecast products, TWEBs cannot cover every instance of hazardous weather. Although the number of TWEB routes is extensive, many areas are not covered. Never extrapolate nor extend a forecast beyond its defined area.

TWEBs are not available through DUATs. However, most NWS forecast offices make this product available on the Internet home page. A list of forecast office home pages is available on the Internet at *www.noaa.gov*.

Significant Weather Progs

Forecasts beyond FA, TWEB, and TAF valid times are available using significant weather prognosis charts. Like the TAF, FA, and TWEB, these charts do not directly forecast icing, but icing can be inferred. Two charts provide forecasts for up to about 48 hours. These charts use standard weather symbols. Charts are available on the Internet at *www.weather.noaa.gov/fax/nwsfax.shtml*.

The 12- to 24-hour Low-Level Significant Weather Prog is issued four times a day, valid at 0000Z, 0600Z, 1200Z, and 1800Z depending on issuance time. The 12- to 24-hour prog consists of four panels. The two upper panels forecast significant weather from the surface to 400 mb (24,000 ft). Forecast weather categories (VFR, MVFR, and IFR) have similar definitions and limitations to the weather depiction chart. Freezing level is represented by short, dashed lines, and the surface 0° isotherm is denoted by a zigzag line and the contraction *SFC*. The lower panels portray the synoptic situation along with areas of significant precipitation. Since the chart is a synoptic depiction, it cannot consider local or localized conditions.

Refer to Fig. 4-1. The left panels are valid at 0000Z on February 25, 1998, the right panels at 1200Z. Forecast freezing levels range from above 12,000 ft in Texas and Florida to below 4000 ft along the Canadian border. Notice that surface temperatures are expected to be at or below freezing over most of the western United States. This could indicate a positive sign for potential SLDs in areas of liquid precipitation. However, any freezing precipitation should be localized; otherwise, the forecast would indicate freezing rain or freezing drizzle.

The marginal VFR to IFR conditions along the southern California coast should have low tops. Why? There are no major weather systems. One would not expect significant icing in this area with the relatively high freezing level.

It is another story over the Rockies. Freezing temperatures at the surface, low freezing levels aloft, marginal VFR to IFR conditions, and widespread precipitation exist. Tops should be relatively high, associated with major weather systems and upslope. One would expect a greater chance of icing in areas where liquid precipitation (rain, rain showers) is forecast at the surface and less icing in areas of solid precipitation (snow). One positive factor for the central and northern Rockies is that with the cold temperatures, any icing layer should be relatively low.

Improving conditions are expected over the Northeast, except for New England. Weather is forecast to be mostly MVFR, except for IFR upslope conditions along the northern Appalachians. Little, if any,

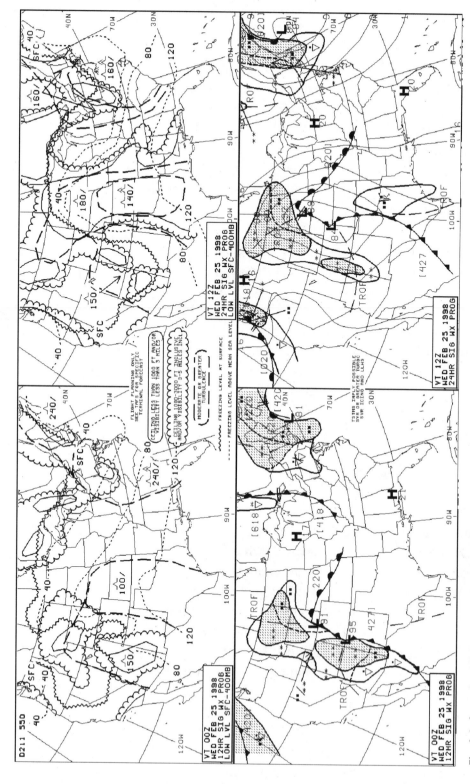

Fig. 4-1. forecasts beyond FA, TWEB, and TAF valid times consult the significant weather prog charts.

overrunning warm air is expected. However, cold arctic air will enter the region associated with the high pressure area over the Great Lakes as it moves eastward. The chart does not indicate any significant lake effect, with relatively light winds associated with the high center. There is a potential for icing over New England, but one would expect a decrease in potential icing moving from west to east.

Strictly a surface prog, the 36- to 48-hour significant weather prog is issued twice daily, valid at 0000Z and 1200Z. The bottom of the chart contains a written narrative of the expected movement of weather systems.

Refer to Fig. 4-2. The left panel is valid at 0000Z on December 2, 1998, the right panel at 1200Z. Based on the time of year and type of precipitation forecast, one can conclude that surface freezing temperature will exist for the northern third of the nation and southward into the Great Basin and southern Rockies. With the weather systems in the west, one would expect significant icing and relatively high tops. One would expect possible SLDs ahead of the cold front in the Pacific Northwest and behind the cold front over the southern Great Basin.

Positive indicators of SLDs are given by a forecast of freezing drizzle in the southwestern portion of the Province of Ontario, Canada, associated with an arctic stationary front. Over the next 12 hours, overrunning warm air from the warm front moving through the Great Lakes will produce SLD icing in southern portions of the provinces of Ontario and Quebec.

In the east, with high pressure dominating, little, if any, icing would be indicated. Nor would one expect icing with the weak cold front in the Midwest. A potential for icing could be expected, however, over Florida, with scattered rain showers and thunderstorms expected. However, without a strong northerly flow and convective activity forecast, one would expect relatively high freezing levels.

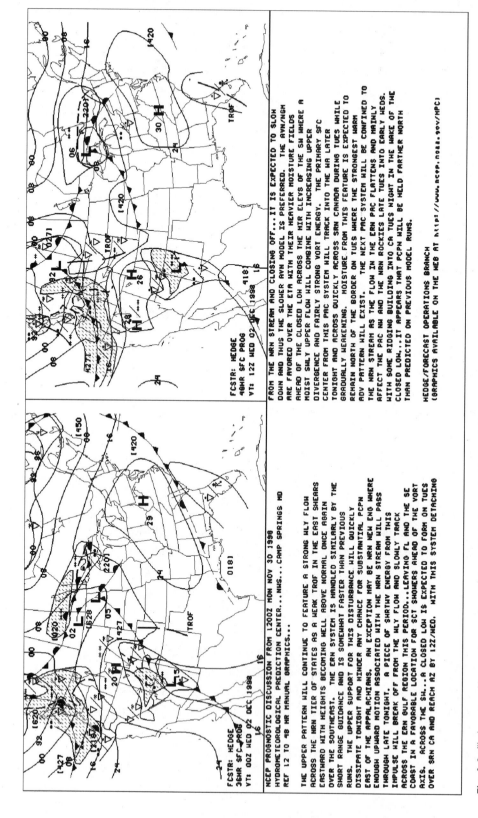

Fig. 4-2. Weather systems with overrunning warm air and freezing precipitation suggest potential supercooled large droplet icing.

Future Developments and Products

This section discusses some of the developments to improve icing forecasts. The following is the FAA's "vision" for icing and its related hazards. Additional information is available on the Internet at *www.faa.gov.aua/ipt_prod/tower/awr/awr.htm.*

The FAA plans to embark on a 7-year research and development program to improvement inflight icing detection and forecasting. The overall theme of the program is to combine information from models and sensors in such a manner as to provide quantifiable improvements in icing diagnoses.

Icing is a cause or factor in numerous fatal aircraft accidents and creates significant disruption to flight operations. Current weather products do not adequately address these dangerous and disruptive events. Avoidance of these events would be possible with improved operationally available high-resolution, accurate forecasts of atmospheric icing conditions.

The goal is an hourly, gridded depiction or forecast of inflight icing based on operational model output combined with real-time sensor data, including icing characteristics (intensity and type). The target probability of detection (POD) is 80 percent. Since at this time there is no valid means to compute false-alarm rate, there is no specific overwarning target until a system is developed as part of this program. The philosophy is to minimize volume warned while not compromising the 80 percent POD. The objective is a combination of an operational model–based icing aviation impact variable (AIV) with supplemental sensor-based guidance to be used as Aviation Weather Center (AWC) AIRMET preparation aids. Model resolution depends on what is operationally available from the National Centers for Environmental Prediction (NCEP); final output resolution and format (text, graphics, or grids) depends on AWC capability development.

The major criticism of traditional icing forecasts is their general nature, often covering entire states. This "shotgun" approach provides little help in fine-tuning flight planning.

FACT

Inflight icing is currently the FAA's top weather research priority.

The AWC needs integrated information on a national scale for producing AIRMETs and SIGMETs. A combination of model outputs and sensor data (surface observations and satellite and radar data) shows promise for achieving this goal. The development of algorithms for raw AIV generation from numerical weather forecast models and for conversion of satellite, radar, and surface observations to icing-relevant fields will supplement the model-based AWC forecaster guidance. Directed research will be pursued into environmental characterization, storm structure, microphysics, sensor design and response, model improvements, and accretion physics needed to ensure needed improvement of these products in the next 2 years and beyond. Emphasis will be placed on operational instruments (GOES 8/9 and NEXRAD), but potential upgrades to systems will be assessed through the use of research facilities with upgraded capabilities.

In the future, pilots and support personnel will have to make more weather-related decisions. Free-flight operations in particular will require more information for the pilot who will be making decisions on where to fly. While the pilots will need more weather information, the communications bandwidth to pilots may not increase significantly for many aircraft. Hence the weather information will need to be tailored to each individual aircraft and its particular situation. Custom forecasts will be required for each aircraft. The only way for custom forecasts to be generated for a large number of aircraft will be to use computer systems that can extract from a larger database the critical information that the specific pilot requires and then format it in an appropriate fashion. A service will be required to provide computer-compatible information to these weather computers serving the aviation community.

The proposed service is the Aviation Gridded Forecast System (AGFS). The AGFS will be an official service of the NWS originating from the Aviation Weather Center (AWC). The database and distribution portion of the AGFS are the Aviation Digital Data Service

(ADDS), which became operational in February 1997. The goal of the AGFS will be to provide accurate, timely, detailed weather observations and forecasts that can be used to derive information for flight planning and operations. The AGFS will require observations and forecasts with details in the tens of miles, with special extent of thousands of miles and vertical resolutions on the order of 1000 ft. Mesoscale models can provide data at these resolutions, but for most weather elements (convection, clouds, icing, turbulence) that affect flight operations, the models are not sufficiently accurate. Even new experimental models of the 1-km resolution range have significant errors and cannot be used directly without human intervention. In the past, human intervention has been in the form of total reformat of model information. The forecaster either writes words or draws graphics that are sent to human users to look at and make decisions. The challenge of the AGFS is to develop technology that will allow a human forecaster to influence the model data rather than doing a total reformat of the data.

Aircraft ground deicing can be costly if not optimized. Costs include the cost of deicing fluid, the cost of deicing crews, and the cost of delays as a result of deicing activities. Decision makers require accurate information about current and forecast weather in order to optimize deicing operations while not compromising safety of flight. A key safety issue with deicing is to prevent the buildup of ice or snow on critical aircraft surfaces prior to takeoff. This type of buildup has been a cause or a factor in numerous aircraft takeoff accidents because of its large effect on aircraft performance, with as little as 1 mm of rough ice reducing lift and increasing drag by 25 percent during the takeoff roll. The failure of these fluids (when snow and ice start to re-form) depends primarily on the anti-icing fluid concentration, liquid equivalent precipitation rate (snowfall or freezing rain), ambient temperature, and wind speed. The current method of determining the snowfall rate using the NWS real-time surface snowfall intensity measurements ($-SN$, SN, $+SN$) based on visibility, however, often can be misleading due to the occurrence of high liquid equivalent snowfall rates.

There is a need to optimize and increase the safety of ground deicing operations (including runway clearing) during snow events at major airports. The approach is to develop an accurate, graphic depiction of the real-time, 30-minute nowcast, and 4- to 10-hour forecasts of precipitation intensity, precipitation type and weather condition, temperature, and wind speed and direction for the 10-km region surrounding an airport. Precipitation type and weather condition include

- Snow

- Freezing rain and freezing drizzle

- Freezing fog

- Ice pellets

- Snow pellets

- Active frost conditions

The FAA has identified these precipitation types and weather conditions, and these time scales (current, 30-minute nowcast, and 4- to 10-hour forecasts) are the most important for ground deicing decision making. The real-time estimates of precipitation rate should be accurate to ± 0.5 mm/h, temperature to ± 0.5°C, wind speed to ± 1 m/s, and wind direction to ± 5°. The 30-minute nowcast of precipitation rate should be of sufficient accuracy to distinguish between three levels of precipitation rate—a light category for rates less than 1.25 mm/h, a moderate category for rates between 1.25 and 2.5 mm/h, and a heavy category for rates greater than 2.5 mm/h. The 30-minute nowcast of temperature should be accurate within 1°C, the wind speed within 2 m/s, and wind direction within 20°. The 4- to 10-hour forecast of precipitation should predict the onset of snow within ± 30 minutes and whether the intensity is to be light, moderate, or heavy. The weather information will be depicted graphically in a user-friendly manner on a computer display. The algorithm and techniques used to generate the graphics will be provided to any commercial vendor interested in providing this product to airlines and airport operators. An important part of the program will be to attract and work with interested vendors. A prototype system will be operationally evaluated at a number of airports affected by winter weather by mutual agreement of the FAA and airlines and airports.

In the Introduction, an excerpt from FAA Administrator Jane Garvey's Aviation Weather Policy was provided. Below is an excerpt from a letter written to the National Weather Association (NWA), in October 1998, by National Weather Service Director John J. Kelly, Jr. It supports the NWS's commitment to leadership and progress in aviation weather forecasting.

> Our aviation customers have expressed serious concerns about NWS aviation services....New aviation products and services and improvements to existing products and services will be designed, validated and implemented....Finally, I reaffirm the NWS dedication to the aviation program, and I look forward to working with you to improve our aviation services.

Progress has been made in the area of improved icing forecasts. Two primary efforts are the stovepipe model and the experimental neural network icing products. Both are available via the Internet.

Pilots are cautioned that these products, other experimental products, and considerable aviation weather data accessible through the Internet do not satisfy weather briefing requirements. Pilots using the Internet exclusively must check with their vendor to ensure that information fulfills their weather briefing obligation. The NWS's disclaimer states:

> **CAUTION**
> Experimental weather forecasting products, and much of Internet weather information, do not satisfy weather briefing requirements.

> Data distribution via the Internet is not considered an operational delivery mechanism by the NWS due to our inability to ensure access to this service, therefore, the information available here shall not be used for flight planning or other operational purposes.

Pilots are further cautioned that the use of experimental products does not relieve them from obtaining official NWS icing products— AIRMET Bulletin, SIGMETs, CWAs.

To begin this discussion, let's examine Figs. 4-3 and 4-4. They graphically and textually contain icing forecasts for November 21, 1998.

In Fig. 4-3, the textual descriptions for both San Francisco and Salt Lake City AIRMET ZULU are identical. They describe a condition of occasional moderate rime or mixed icing in clouds and precipitation. Therefore, one needs to consider possible SLD icing.

Figure 4-4 illustrates a situation where the forecaster has warned of isolated severe rime or mixed icing in cloud and precipitation below 10,000 ft. Icing has become serious enough to issue a SIGMET (gray-shaded area in Fig. 4-4) for essentially the same area described as isolated severe in the AIRMET but specifically below 8000 ft. SLD encounters can be expected in these areas.

To the west, Chicago AIRMET ZULU advertises a less serious condition—light to occasional moderate rime icing. SLD icing is not expected. Note that the forecaster describes the condition as shifting eastward over the area during the period—between 1500Z and 2100Z—and to continue eastward through 0300Z. This is a good example of perceived overforecasting. Rather than affecting the whole area, the hazard will move through the area during the period. Unfortunately, FSS briefers are not specifically trained on how to interpret and translate this advisory and typically will not relay the "shifting eastward" portion of the AIRMET. Who's job is it to determine if the advisory is applicable to a specific flight? That's right, the pilot.

There is an excellent correlation between weather advisories and experimental products. In most instances, the models support the weather advisories. In a few cases, the models provide additional insight into the advisories. Typically, the model's spatial coverage correlates very well with the weather advisories; in some cases it does not. The reasons will become apparent.

Some pilots have touted these products as providing an icing panacea. I do not share this view. They are simply another piece of the icing puzzle and must be used along with all other weather reports and forecasts for an intelligent flight planning decision.

One additional note. Most icing models and studies use $-2°C$ as the starting point for icing. This is due to the aerodynamic warming effects of compression and friction. This, along with the fact that icing intensities vary throughout the cloud structure, helps explain why icing levels and the forecast freezing level do not always match in the AIRMET Bulletin.

With the preceding as a background, let us continue the discussion of experimental products. Note that the time frames are the same as those contained in Figs. 4-3 and 4-4.

Experimental Neural Network Icing Products (AWC)

Experimental neural network icing products have been developed by the Experimental Forecast Facility (EFF) at the NWS's Aviation Weather Center (*www.awc-kc.noaa.gov*). They consist of an initial analysis and graphic forecasts out to about 12 hours.

According to Don McCann, EFF meteorologist,

The patterns of the conditions for significant icing (temperature, relative humidity, and slight convective potential) are numerous and quite complex. They do not lend themselves to typical input/output prediction schemes. An artificial intelligence tool that does very well at pattern recognition is a neural network. A neural network is a simple version of a mammalian brain. A neural network is "taught" to recognize input data patterns and to identify with what conditions these input patterns are associated. A neural network is pure rote learning, like a grade-schooler learning multiplication tables. Neural networks work well in complex pattern recognition because they are very nonlinear. That means that they can handle situations which cannot be expressed in a straightforward logical thought-stream.

The maps of icing intensity available below are composites of the output of two neural networks taught to predict icing intensity from input data of temperature, relative humidity, and convective potential from the Rapid Update Cycle (RUC) model. Actual output to the Aviation Weather Center forecasters is in layers approximately 1000

ICING AIRMETs for after 981121/1445

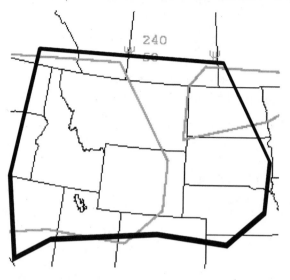

SFOZ WA 211445
AIRMET ZULU UPDT 2 FOR ICE AND FRZLVL VALID UNTIL 212100

.
AIRMET ICE...WA OR CA AND CSTL WTRS ID MT WY NV UT CO
FROM YXH TO GCC TO LAR TO GJT TO ELY TO SFO TO 150SW SFO TO
120W FOT TO 120W TOU TO YXH
OCNL MOD RIME OR MXD ICGICIP BTN 050-080 AND FL240 OVR NRN AND
CNTRL PTNS AREA AND 100-120 AND FL240 OVR SRN PTNS AREA. CONDS
CONTG BYD 21Z THRU 03Z.

SLCZ WA 211445
AIRMET ZULU UPDT 2 FOR ICE AND FRZLVL VALID UNTIL 212100

Fig. 4-3. It is often helpful to plot the AIRMET Bulletin and SIGMETs on graphs; one source is the AWC Internet home page.

ft thick. The output values range from zero to six, with zero representing no icing and six severe icing. A two is light icing and a four is moderate icing. The contours begin at the two level. While fours are very common, a five (moderate to severe icing) is rare. Because of the contouring routine, you will never see a six.

This is experimental output. The Experimental Forecast Facility at the Aviation Weather Center is evaluating the output to see if there are any flaws. Anyone using this product as a flight-briefing aid should always consult the latest Aviation Weather Center icing advisories.

ICING AIRMETs for after 981121/1445

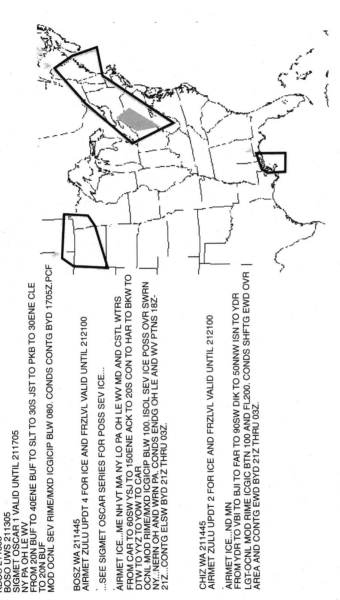

KBOS 211305
BOSO UWS 211305
SIGMET OSCAR 1 VALID UNTIL 211705
NY PA OH LE WV
FROM 20N BUF TO 40ENE BUF TO SLT TO 30S JST TO PKB TO 30ENE CLE
TO 20N BUF
MOD OCNL SEV RIME/MXD ICGICIP BLW 080. CONDS CONTG BYD 1705Z.PCF

BOSZ WA 211445
AIRMET ZULU UPDT 4 FOR ICE AND FRZLVL VALID UNTIL 212100

...SEE SIGMET OSCAR SERIES FOR POSS SEV ICE...

AIRMET ICE...ME NH VT MA NY LO PA OH LE WV MD AND CSTL WTRS
FROM CAR TO 60SW YSJ TO 150ENE ACK TO 20S CON TO HAR TO BKW TO
DTW TO YYZ TO YOW TO CAR
OCNL MOD RIME/MXD ICGICIP BLW 100. ISOL SEV ICE POSS OVR SWRN
NY...NERN OH AND WRN PA. CONDS ENDG OH LE AND WV PTNS 18Z-
21Z...CONTG ELSW BYD 21Z THRU 03Z.

CHIZ WA 211445
AIRMET ZULU UPDT 2 FOR ICE AND FRZLVL VALID UNTIL 212100

AIRMET ICE...ND MN
FROM YDR TO VBI TO BJI TO FAR TO 90SW DIK TO 50NNW ISN TO YDR
LGT-OCNL MOD RIME ICGIC BTN 100 AND FL200. CONDS SHFTG EWD OVR
AREA AND CONTG EWD BYD 21Z THRU 03Z.

Fig. 4–4. SIGMET OSCAR has been superimposed (the gray area) on this AIRMET ZULU graphic.

The following are neural network icing product output values:

- 0: No icing

- 1: No icing to light icing

- 2: Light icing

- 3: Light to moderate icing

- 4: Moderate icing

- 5: Moderate to severe icing

- 6: Severe icing

Figures 4-5 through 4-8 represent the experimental neural network icing products. Figure 4-5 is the surface to 6000-ft composite. The product correlates well with the icing advisories. In the Pacific Northwest, light to moderate icing (2–4) covers the northern portion of the icing advisory. Why not the south? The base of icing in this area is forecast between 10,000 and 12,000 ft—above the height of the composite.

Again, in the north central and northeastern United States, agreement on coverage and intensity is consistent, e.g., threes in North Dakota and Minnesota and fours and isolated fives in the Northeast. Why not an advisory in Utah and Arizona? No clouds. The lack of ability to consider cloud cover is a limitation of this product.

Figure 4-6 shows the 6000- to 14,000-ft composite. Notice how at this level in the Northwest the area of significant icing has moved south. Light to moderate and moderate icing is forecast in western Dakota. Recall how the forecaster advised that conditions would be shifting eastward during the period. The Northeast also correlates well for both area and intensity. There is an area of light to moderate and isolated moderate icing, for this level, along the Gulf Coast.

Moving on to Fig. 4-7, the 14,000- to 30,000-ft composite, the Northwest and North Central areas correspond well with coverage and

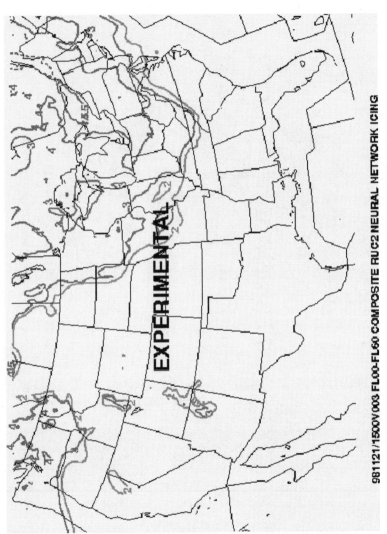

9B1121/1500V003 FL00-FL60 COMPOSITE RUC2 NEURAL NETWORK ICING

Fig. 4-5. Pilots are cautioned about the use of experimental products and directed to always consult current weather advisories for operational planning purposes.

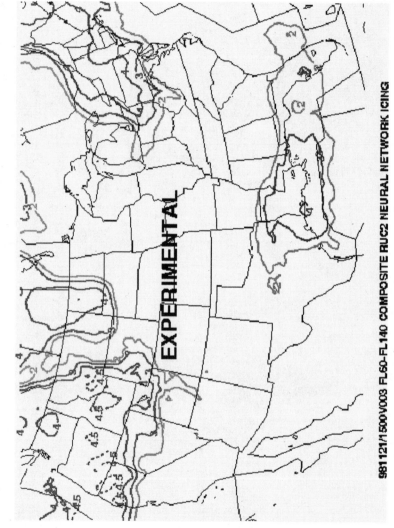

981121/1500V003 FL60-FL140 COMPOSITE RUC2 NEURAL NETWORK ICING

Fig. 4-6. Spatial coverage of the icing event along the Gulf Coast is greatly exaggerated in this experimental product.

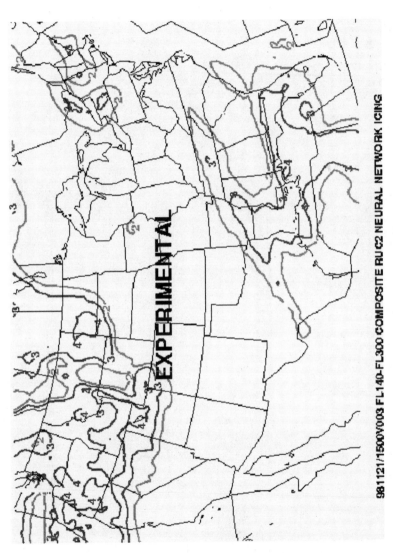

981121/1500V0003 FL140-FL300 COMPOSITE RUC2 NEURAL NETWORK ICING

Fig. 4-7. Icing intensities in the Northwest and Northeast are consistent with the weather advisories.

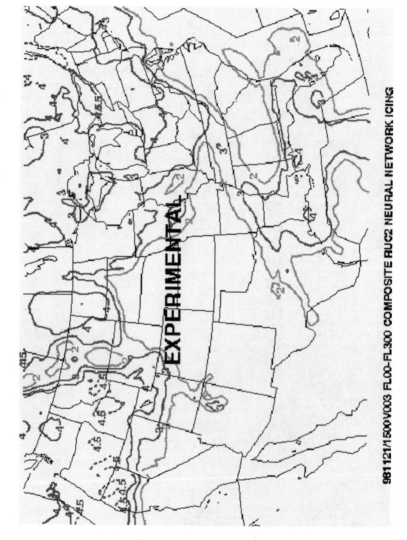

Fig. 4-8. Icing intensity between this surface to 30,000-ft icing composite and the weather advisories show consistent agreement.

intensity. Icing is not expected at this level in the Northeast, due to much colder temperatures aloft. From the weather advisories, icing tops are at 24,000 ft in the Northwest, 20,000 ft in the North Central area, and below 10,000 ft in the Northeast. The model has some serious problems with coverage along the Gulf Coast, although the highest intensities correlate well with the advisory.

Figure 4-8, an all-altitudes composite (surface to 30,000 ft), again correlates well with the weather advisories. Notice, however, that the product overestimates spatial coverage. Otherwise, intensities correlate very well. From the discussion thus far, we should all be able to see that although useful in a general sense, this product is not a substitute for current weather advisories. Icing intensity correlated very well with the advisories. Aerial coverage correlated well in the northern part of the country but was relatively poor in the south. The reasons for the discrepancies? The model cannot take all the factors available to the forecaster into consideration. Future refinements of the model may be able to accommodate these factors.

The Stovepipe Model (NCAR)

The following discussion and charts were obtained from the National Center for Atmospheric Research (NCAR), developed by Ben Bernstein and Frank McDonough at the NCAR Research Applications Program, Boulder, Colorado (*www.ucar.edu/wx.html*). I asked Ben, why "stovepipe"? Ben responded:

The Stovepipe algorithm uses data from surface observations within a 100-km kilometers (150 km in the western United States) radius around an RUC grid point. [The Rapid Update Cycle (RUC) is a high-speed computer model that updates every 3 hours, designed for short-term forecasting.] That radius can be thought of as the "belly" of a "pot-belly stove." Then, the surface observation information is combined with the column of temperature and relative humidity data from the RUC grid point, which is the "stovepipe." Make sense?

Yes, Ben, I think it does.

The stovepipe icing algorithm uses information available from surface observations and three-dimensional gridded fields of temperature, relative humidity, and geopotential height—a measure of atmospheric energy—to create a three-dimensional diagnostic of icing conditions. Since new surface observations are available every hour, the algorithm is run hourly.

The algorithm is based on research that has shown that nearly all pilot reports of icing occur in regions of precipitation or overcast cloud conditions. Research also has shown that an unusually high number of moderate or greater-intensity PIREPs that reported mixed or clear icing occurred in these areas. When these conditions are observed at the surface, precipitation-sized supercooled large drops (SLDs) will exist through some depth above the surface.

This information has been employed by the stovepipe algorithm, which looks for surface observations of freezing drizzle, freezing rain, and ice pellets within 100 km of each rapid update cycle (RUC) model grid point. If such conditions are found, the algorithm searches for a narrow range of temperature and relative humidity associated with the moderate or greater-intensity PIREPs. These locations are where SLD conditions are likely to exist. If these conditions are not found, the algorithm checks for any other precipitation or overcast sky conditions. If those conditions are found, the algorithm applies a slightly different temperature and relative humidity range associated with PIREP occurrences of less serious icing.

The algorithm improves greatly on those which use temperature and relative humidity blindly, since the stovepipe model will only diagnose icing in those locations where at least overcast conditions exist—an advantage over the neural network. Basic temperature–dew–point algorithms use thresholds of relative humidity that sometimes extend well below 70 percent and can predict icing in the absence of clouds.

Experience from aircraft icing field projects, case studies, past algorithm development, and real-time forecasting has shown that the potential for icing is quite different for different weather scenarios. The occurrence of these different scenarios can be identified through the use of data from satellites, surface observations, radar, and three-dimensional model grids.

For each scenario, a mathematical equation has been developed to combine information from the four data sources to estimate the potential for icing at each grid point. These mathematical equations are designed to be physically consistent with the microphysical scenario that is likely to exist within each column. The final result is a floating-point value of 0 to 100 for both the icing and SLD potential, as well as icing type at each grid point in the model. Larger values of icing and SLD potential indicate a greater likelihood that those conditions will exist. The icing-type field is a first guess, simply based on temperature at each model grid point, with warmer temperatures implying a greater likelihood of clear/glaze icing and colder temperatures implying a greater likelihood of rime icing.

The SLD potential field is designed to identify locations where precipitation-sized supercooled water drops are most likely to exist. Similar to the icing field, this potential is based on a combination of information from surface observations, satellite, radar, and RUC model data. Essentially, this portion of the algorithm looks for the following situations:

1. *Freezing precipitation below a warm nose*—a prominent northward bulge of relatively warm air. When a warm nose has been identified in the RUC model data and liquid or freezing precipitation (*FZDZ, FZRA, PL, DZ,* or *RA*) is occurring at the surface, SLDs are likely to be present at any altitude below the warm nose where temperatures are subfreezing.

2. *Freezing precipitation without a warm nose.* The mere occurrence of freezing drizzle, freezing rain, or ice pellets at the surface is an absolute indicator of the existence of SLDs aloft. Without the presence of a warm nose, the SLD must be forming via the collision-coalescence process. It typically will extend from the surface to the top of the lowest cloud deck.

3. *Nonfreezing liquid precipitation (drizzle and rain) without a warm nose but with warm cloud-top temperatures.* Both drizzle and rain can form through a collision-coalescence process or from melting snow that reaches the ground as liquid precipitation. The trick is to know which of these processes are occurring.

Cloud-top temperature data from satellites are used to help answer these questions. When cloud tops are rather warm (approximately greater than −12°C, the likelihood becomes relatively high that the entire cloud deck is made up of liquid water. On the flip side, when cloud-top temperatures become relatively cold, the likelihood increases that the rain or drizzle is forming from snow melting as it falls through the freezing level. Thus SLD potential will only be indicated in this situation when warm cloud tops are present, and the potential increases as the cloud-top temperature increases.

4. *Above the warm nose when liquid or freezing precipitation is occurring at the surface and cloud-top temperatures are warm.* Field program and case-study data have revealed that SLDs can exist above the warm nose in situations where cloud-top temperatures are quite warm (approximately greater than −15°C). This is so because the layer above the warm nose is one where overrunning is often occurring, which is conducive to the formation of SLDs only when the cloud is primarily in a liquid state. The phase is very important because gradual lift in a cloud that is primarily ice is not likely to yield SLDs. Even if the cloud is primarily liquid in phase, either SLDs or cloud drop icing will form.

There are five plots given for the observation-based stovepipe algorithm.

1. *Observation-based stovepipe algorithm—surface projection.* This plot shows all the locations where icing has been diagnosed at some level in the model. If red is shown, then SLD was diagnosed at some level, and general icing probably was found at other levels. If blue is shown, then general icing was diagnosed at some level, but SLD was not diagnosed at any level in the model sounding. The locations of all pilot reports of icing are superimposed on this plot so that a real-time

evaluation of the algorithm's performance can be done at a glance. A description of the plotted PIREPs is given in the lower left corner of the plot. Real-time verification of the icing algorithm is given in the lower right portion of the plot, including what portion of the icing PIREPs were diagnosed correctly by each portion of the algorithm and one measure of how efficient the algorithm was for that hour.

2. *Observation-based stovepipe—SLD icing bases.* Here, the altitude of the base of the SLD layer at each model grid point is color coded by range. The ranges are indicated at the bottom left portion of the plot. The altitudes of all PIREPs of moderate or greater intensity icing are given in thousands of ft. These PIREPs are a fairly good indicator of aircraft encounters with SLDs.

3. *Observation-based stovepipe—SLD icing tops.* Here, the altitude of the top of the SLD layer at each model grid point is color coded by range. The ranges are indicated at the bottom left portion of the plot. The altitudes of all moderate or greater intensity icing that was mixed or clear PIREPs are given in thousands of ft. These PIREPs are a fairly good indicator of aircraft encounters with SLDs.

4. *Observation-based stovepipe—general icing bases.* Here, the altitude of the base of the general icing layer at each model grid point is color coded by range. The ranges are indicated at the bottom left portion of the plot. The altitudes of all PIREPs of moderate or greater intensity icing are given in thousands of ft.

5. *Observation-based stovepipe—general icing tops.* Here, the altitude of the top of the SLD layer at each model grid point is color coded by range. The ranges are indicated at the bottom left portion of the plot. The altitudes of all PIREPs of moderate or greater intensity icing are given in thousands of ft.

Figures 4-9 through 4-15 represent the stovepipe icing product. Their valid time corresponds to the icing advisories in Figs. 4-3 and 4-4.

Figure 4-9, the surface projection, plots locations where icing has been diagnosed at some level. The model correlates well with the general icing areas in the weather advisories, although, like the neural

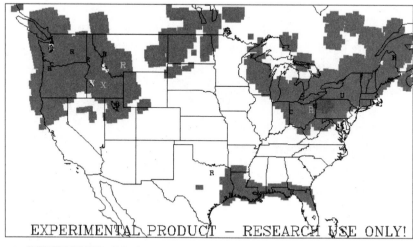

Map of STOVEPIPE ALGORITHM 981121 — 16 Z
Radius of influence was 100 km
SURFACE PROJECTION
● Large Supercooled Drops Likely
● General Icing
● No Data For These Points

EXPERIMENTAL PRODUCT — RESEARCH USE ONLY!

PIREPS FROM 98112116 to 98112116
ICING PIREP INDICATORS

C = Clear Icing
X = Mixed Icing
R = Rime Icing
U = Unknown — No Type Indicated
MOD/SEV to SEV = Large/Orange
LGT/MOD to MOD = Medium/Yellow
TRC to LGT = Small/Black

Fig. 4-9. The stovepipe surface projection plots locations where icing has been diagnosed at some level.

network, it overestimates spatial coverage. The model very accurately depicts the location of SLD probability—the SIGMET area in Fig. 4-4.

In Fig. 4-10 the bases of SLDs are depicted. Again, correlation with the weather advisories is good. The model predicts bases between 1000 and 3000 ft MSL. Taking into account terrain in the depicted area, the SLD base is at or near the surface. This is consistent with the SIGMET. In this area, a pilot could expect to carry any ice accumulations all the way to the ground! This is a significant hazard for all aircraft and must be considered carefully before flight.

Map of STOVEPIPE ALGORITHM 981121 – 16 Z

Radius of influence was 100 km

LARGE DROPLET BASE

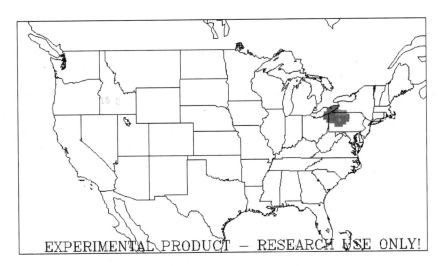

EXPERIMENTAL PRODUCT – RESEARCH USE ONLY!

- BELOW 1000 FT MSL
- 1000 TO 2000 FT MSL
- 2000 TO 3000 FT MSL
- 3000 TO 4000 FT MSL
- 4000 TO 5000 FT MSL
- ABOVE 5000 FT MSL
- No Data For These Points

Fig. 4-10. This graphic depicts supercooled large drop bases, which are consistent with the AIRMET Bulletin and SIGMET OSCAR.

The top of the SLD event is presented in Fig. 4-11. Tops of the large supercooled drops are expected to be between 7000 and 10,000 ft MSL, again consistent with the SIGMET. This information would be of considerable value to a pilot planning a flight through this area. Even with ice protection equipment, a pilot would want to stay above 10,000 ft.

Figure 4-12 depicts general icing bases. In the Northwest, bases are expected to be between 4000 and 7000 ft MSL in northern portions and at about 12,000 ft in southern portions. In the north central United States, bases are predicted to be between about 5000 and

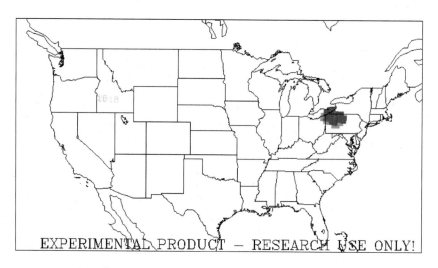

Map of STOVEPIPE ALGORITHM 981121 — 16 Z

Radius of influence was 100 km

LARGE DROPLET TOP

EXPERIMENTAL PRODUCT — RESEARCH USE ONLY!

- ● BELOW 1000 FT MSL
- ● 1000 TO 4000 FT MSL
- ● 4000 TO 7000 FT MSL
- ● 7000 TO 10000 FT MSL
- ● 10000 TO 15000 FT MSL
- ● ABOVE 15000 FT MSL
- ● No Data For These Points

Fig. 4-11. Supercooled large drop tops and location are in agreement with the weather advisories.

10,000 ft. And in the Northeast, bases are forecast to be between 1000 and 3000 ft—this translates to the surface in much of this area. There is good correlation with the weather advisories.

Figures 4-13 through 4-15 predict potential icing at various levels. The stovepipe model forecasts icing from 3000 to 24,000 ft, at 3000-ft intervals. This discussion will review the 6000-, 12,000-, and 18,000-ft level only.

In Fig. 4-13, the model predicts isolated areas of high probability (80 to 100 percent) of icing in the Northwest at the 6000-ft level.

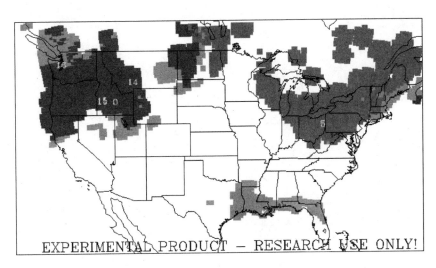

Map of STOVEPIPE ALGORITHM 981121 – 16 Z

Radius of influence was 100 km

----ICING BASES---

EXPERIMENTAL PRODUCT – RESEARCH USE ONLY!

- ● 0 TO 1000 FT MSL
- ● 1000 TO 3000 FT MSL
- ● 3000 TO 5000 FT MSL
- ● 5000 TO 10000 FT MSL
- ● 10000 TO 15000 FT MSL
- ● ABOVE 15000 FT MSL
- ● No Data For These Points

PIREP BASES IN 1000s OF FT

MOD/SEV to SEV = Large/Orange

LGT/MOD to MOD = Medium/Yellow

Fig. 4-12. Like the neural network, this stovepipe model exaggerates spatial coverage of the icing event along the Gulf Coast.

Scattered area of moderate probability (30 to 70 percent) and low probability in the southern portions (<30 percent). No icing potential exists at this level in the north central United States. Widespread high probability exists in the Northeast. There is good agreement with the weather advisories.

Figure 4-14 depicts the 12,000-ft level. There are isolated areas of moderate to high probability over northern California, northern Utah, and southeastern Idaho. Otherwise, scattered areas of moderate probability exist over most of the area. Generally, the same conditions are expected over the north central United States. Over the Northeast,

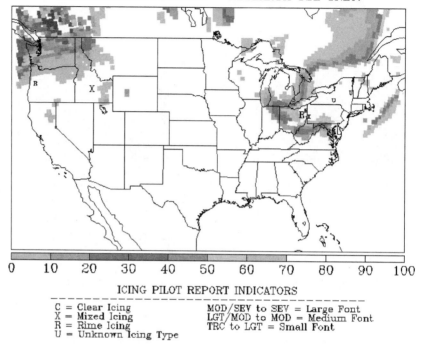

INTEGRATED ICING ALGORITHM 981121 −16 Z
POTENTIAL FOR ICING AT 6000 FT
EXPERIMENTAL PRODUCT − RESEARCH USE ONLY!

ICING PILOT REPORT INDICATORS

C = Clear Icing   MOD/SEV to SEV = Large Font
X = Mixed Icing   LGT/MOD to MOD = Medium Font
R = Rime Icing   TRC to LGT = Small Font
U = Unknown Icing Type

Fig. 4-13. The stovepipe model icing potential at 6000 ft agrees very well in both coverage and intensity with the weather advisories.

widespread areas of low to moderate probability are depicted. Now, at this level, moderate probability is forecast for the Gulf Coast. These predictions, except for some spatial coverage difference, correlate very well with the weather advisories.

Whatever may be the progress of the sciences, never will observers who are trustworthy and careful of their reputations venture to forecast the state of the weather. [Dominique Argo, French astronomer, 1786–1853].

Reviewing Fig. 4-15, we see that there is no icing potential at the 18,000-ft level for northern Washington state. It would seem that the icing tops in AIRMET ZULU were a "strategic forecast." This emphasizes the point that pilots must evaluate all available data for flight planning. They would want to review current conditions, the area forecast, and winds and temperatures aloft. With all information evaluated, and a knowledge of icing, a pilot should be able to make a sound

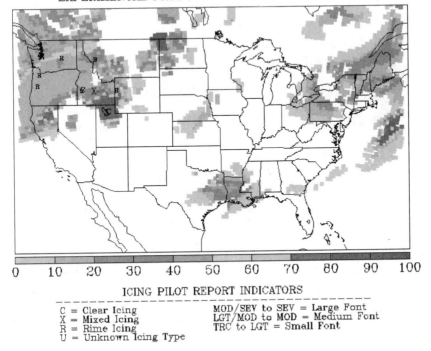

INTEGRATED ICING ALGORITHM 981121 −16 Z
POTENTIAL FOR ICING AT 12000 FT
EXPERIMENTAL PRODUCT − RESEARCH USE ONLY!

0 10 20 30 40 50 60 70 80 90 100

ICING PILOT REPORT INDICATORS

C = Clear Icing
X = Mixed Icing
R = Rime Icing
U = Unknown Icing Type

MOD/SEV to SEV = Large Font
LGT/MOD to MOD = Medium Font
TRC to LGT = Small Font

Fig. 4-14. Compared with the weather advisories, the stovepipe model icing potential at 12,000 ft correlates well for intensity, but coverage is overestimated.

go-no-go decision. No icing potential exists for the Northeast, which is what one would expect. A light to moderate potential for icing exists over the southeast United States. This corresponds with the AIRMET along the Gulf Coast, but the model's spatial coverage, similar to the neural network icing product, is poor.

ICING MYTH

Forecasts for icing are always exaggerated.

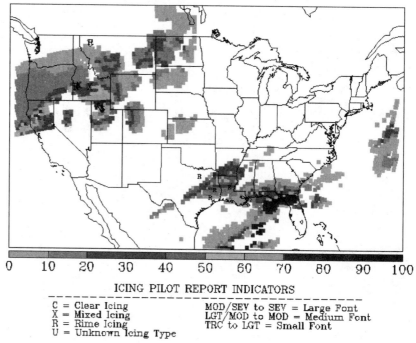

INTEGRATED ICING ALGORITHM 981121 −16 Z
POTENTIAL FOR ICING AT 18000 FT
EXPERIMENTAL PRODUCT − RESEARCH USE ONLY!

ICING PILOT REPORT INDICATORS

C = Clear Icing   MOD/SEV to SEV = Large Font
X = Mixed Icing   LGT/MOD to MOD = Medium Font
R = Rime Icing   TRC to LGT = Small Font
U = Unknown Icing Type

Fig. 4-15. Again, at 18,000 ft, the stovepipe model icing potential intensity agrees with the weather advisories.

CHAPTER 5

Flight Planning Strategies

At this point you should have a sound understanding of the effects of icing and how it is reported and forecast. Armed with this knowledge, it is time to move on to flight planning and cockpit strategies. As you will see, these subjects are often interrelated. Therefore, even though the focus of this chapter is primarily flight planning, occasionally topics that apply to en route operations will be discussed.

Let us first look at the regulations as they apply to ice protection certification and flight rules. This is followed by flight planning. Flight planning consists of evaluating all elements physically related to the flight, such as terrain, altitudes, and the environment.

With an evaluation of terrain and altitudes complete, the next step is assessment of the environment—weather, personal minimums, and alternates. The first part of the environmental evaluation involves obtaining the "complete picture" through a preflight weather briefing. This knowledge is then applied to personal minimums and available alternates. An assessment of terrain, altitudes, and environment is the first step in the go-no go decision.

Unlike being a "little pregnant," there is some middle ground when it comes to the weather. For example, one can plan a flight in stages, landing short of the ultimate destination. One can take a look at the weather, although there are two caveats to this option. First, it is important to know when to abandon the plan. When it is not meant to be, it is not meant to be. One must know when to call it a day. Second, it is necessary to have an alternate plan or two (plan B, plan C, plan n). More about this later in this chapter and in Chapter 6, "Cockpit Strategies."

CASE STUDY

I was on a flight from Amarillo, Texas, to Albuquerque, New Mexico. The weather was good through Tucumcari but deteriorated between Tucumcari and Albuquerque. Passing Tucumcari, I checked with Flight Watch and received the bad news. The weather ahead was IFR to MVFR. The plan was to fly direct, via the Anton Chico and Otto VORs. With the deteriorating weather ahead, I decided to go IFR—I Follow Roads. With little in the way of landmarks, low ceilings, and even lower visibilities, the safest option was to follow I-40. The terrain and clouds merged about 20 mi west of Santa Rosa, New Mexico. It was afternoon, and I had been flying for about 4 hours. With night approaching, poor weather, and fatigue a factor, the only viable option was to return and land at Santa Rosa.

Back to flight planning. With a go decision, the next step is preparation. Besides the aircraft, its airframe, engine, and equipment, pilots must consider themselves. Are they fit for flight? This means certified, current, and physically and mentally prepared, and do not forget the passengers. Are they physically prepared for the flight? If icing is expected, it must be cold. What would happen in the event of an emergency landing? Is everyone prepared to survive a day or two in the open? I will talk about this and other survival issues later.

"Safe does not mean risk-free."

As observed by the U.S. Supreme Court: "Safe does not mean risk-free." This leads to risk assessment and management. As pilots, we know about certification and operating rules. We have learned to get a "complete picture" of the weather. We know our aircraft and its equipment and ourselves and our passengers. How do we decide if a particular flight is safe? How do we assess the risk and manage that risk? I have discussed various scenarios thus far. I will specifically address these questions in the final section of this chapter.

Regulations and Certification for Flight in Icing

I touched on certification requirements and how they evolved in the ice protection equipment section of Chapter 2. I also touched on regulations. Now let me put this knowledge together. How does this apply to flight planning strategies? Before a pilot can develop a sound plan, he or she must know the options. With an icing hazard, this means what, if any, ice is the aircraft capable of handling and what, if any, regulations apply.

For an aircraft to be approved for flight into icing conditions, the aircraft must be equipped with systems that adequately protect various components. There are two regulatory references to ice protection: the application to aircraft type certification in 14 CFR parts 23 and 25 and the operating rules contained in 14 CFR parts 91, 121, and 135. Translation: The aircraft must be certified by the FAA for flight in icing in accordance with the airworthiness standards in the regulations (parts 23 through 29—light and commuter aircraft and transport aircraft); additionally, the aircraft must be operated in accordance with the flight rules described in the regulations (parts 91, 121, and 135—general operating rules and air carrier and air taxi operating rules).

With regard to ice protection, the rules require an analysis to establish the adequacy of the ice protection system for the various components, based on operational needs of that aircraft. In addition, tests of the ice protection system must be conducted to demonstrate that the aircraft is capable of operating safely in continuous maximum and intermittent maximum icing conditions, as described in the regulations. The regulations specify liquid water content, droplet size, and duration of flight in icing conditions required for certification. The airplane flight manual (AFM) lists the equipment required to be installed and operating and contains recommended procedures for the use of the equipment— recall the section on the Commander 114 in Chapter 2.

> **FACT**
> An aircraft equipped with some types of deice or anti-ice systems may not be approved for flight into icing conditions.

It is essential to understand that an aircraft equipped with some types of deice or anti-ice systems may not be approved for flight into icing conditions. Pilots who fly aircraft in ice and who are not specifically approved for such flights must have the "right stuff" because in every sense of the word they become test pilots. To be approved for such flight, the aircraft must be specifically certificated to operate in icing conditions.

In addition, it is imperative for pilots to understand that the certification standards provide protection for most atmospheric conditions encountered but not for freezing rain or freezing drizzle or for conditions with a mixture of supercooled droplets and snow or ice particles. Some airfoils are degraded by even a thin accumulation of ice aft of the ice protection equipment, which occurs commonly in these conditions.

Operating rules permit flight into specified icing conditions provided that the aircraft has functioning ice protection equipment. Some aircraft with partial installations of deicing or anti-icing equipment do not meet the certification or operating requirements for flight into icing conditions. Such installations are approved because it has been demonstrated that the equipment does not adversely affect the aircraft's structure, systems, flight characteristics, or performance. In such cases, the AFM or other approved material must explain the appropriate operating procedure for the partial ice protection equipment and contain a clear statement that the aircraft is not approved for flight into known icing.

Let's start at the beginning. Pilots of large and turbine-powered multiengine airplanes are prohibited from takeoff with frost, snow, or ice adhering to a propeller, windshield, or powerplant installation or to an airspeed, altimeter, rate of climb, or flight attitude instrument system. Takeoff is prohibited with snow or ice adhering to the wings or stabilizing or control surfaces or any frost adhering to the wings or stabilizing or control surfaces—unless that frost has be polished to make it smooth. A similar rule applies to air taxi and air carrier operations.

CASE STUDY
Ice on the airframe was the case with a Cessna Caravan. The aircraft was parked outside where freezing rain fell. The next day the pilot removed about 80 percent of the snow covering the wings, which left a coarse layer of ice about 3/16 in thick. The aircraft crashed after takeoff. The probable cause was determined to be the pilot's failure to remove ice from the airframe prior to takeoff.

Figure 5-1 shows an area of frost on the wing of a Piper Arrow. The sun has begun to melt the ice on the wing tip. All frost must be removed before takeoff. The solution to this problem is to hanger the aircraft, arrange for a deicing service, or plan the departure later in the day when the sun has melted the ice.

Pilots are permitted to fly IFR into known or forecast moderate icing conditions or VFR into known light or moderate icing conditions— freezing drizzle, ice pellets, snow—with approved ice protection equipment. No one, even with approved ice protection equipment, is permitted to fly into known or forecast severe icing conditions!

However, if current weather reports and briefing information indicate that forecast icing conditions that would have prohibited the flight will not be encountered because of changed weather conditions since the forecast, the flight is permitted. Huh? (This was translated and summarized from the "bureaucracies" of the regulations.) This is a loophole put in the regulations because the federal regulators know that forecasts are not always correct. Ha! What does it mean?

Fig. 5-1. The solution to the frost hazard is to hanger the aircraft, arrange for a deicing service, or plan a departure later in the day when the sun has melted the ice.

On a particular day, a fast-moving cold front was forecast to move into central California. The forecast contained the usual icing advisories. However, as it turned out, the front stalled just off the coast. An icing advisory extended all the way down to Bakersfield in the southern San Joaquin Valley. It became apparent that the icing forecast was a "bust." In this case, "current weather reports and briefing information indicated that the forecast icing conditions, which would have prohibited the flight, would not be encountered because of changed weather conditions since the forecast." I talked with the forecaster about the mountain obscuration advisory; these conditions also had not developed. The forecaster amended that advisory, and I assumed that the icing advisory also would be updated. Silly me! As a briefer, I discontinued issuing the advisory, and the next issuance, several hours later, deleted icing south of the Bay Area. I am pleased to say this is the exception rather than the rule.

Remember in the previous discussion of PIREPs how personal, social, and/or political comments are sometimes transmitted? On the occasion just mentioned, the following PIREP appeared:

UA /OV BFL/TM 1800/FL100/TP BE90/SK SKC/IC MOD

Definitely a case of "clear air icing!" It reflects the frustration of pilots and flight service station (FSS) controllers over a clearly incorrect forecast.

A SUMMATION OF CANADIAN ICING REGULATIONS

No person shall conduct a takeoff or continue a flight in an aircraft where icing conditions are reported to exist or are forecast to be encountered along the route of flight unless the pilot-in-command determines that the aircraft is adequately equipped to operate in icing conditions in accordance with the standards of airworthiness under which the type certificate for that aircraft was issued or current weather reports or pilot reports indicate that icing conditions no longer exist.

No person may dispatch or release an aircraft, take off an aircraft, continue to operate an aircraft en route, or land an aircraft when icing conditions are expected or encountered if, in the opinion of the pilot-in-command, the icing conditions might adversely affect the safety of the flight.

No person shall operate an aircraft in icing conditions at night unless the aircraft is equipped with a means to illuminate or otherwise detect the formation of ice.

Canadian and U.S. regulations are essentially the same. Does this mean that it is safe to fly small, reciprocating-engine aircraft with frost, snow, or ice adhering to the aircraft or fly noncertified aircraft into reported or forecast icing? No! Well, there's no rule specifically against it. Wanna bet? How about the good old "careless or reckless operation?" This subject was broached before. Just because there is no specific prohibition does not mean that an operation is safe, recommended, or without possible enforcement action. Nor does icing certification permit flight in all icing conditions. It is the responsibility of the pilot-in-command to determine if the pilot, the aircraft, and the weather are suitable to conduct a safe flight.

The Weather Briefing

Little need existed for meteorologic information in the early days of aviation because all flights were local. Nor was aeronautical information, Notices to Airmen (NOTAMs), necessary because pilots departed and landed at the same field, assuming that the engine did not quit. By the spring of 1918, the U.S. Post Office Department began working on a transcontinental airmail route. A combination rail-air route between New York and Chicago was established by July and a month later extended to San Francisco. Authorization was granted in August 1920 for the establishment of 17 Airmail Radio Stations. Personnel originally were to load and unload mail. However, as traffic increased, the need for weather information became apparent. Airmail radio personnel soon began taking weather observations and developing forecasts. The information was relayed via radio telegraph to adjacent stations. Inflight weather reports were heavily relied on for the weather briefing.

Postal personnel soon became involved in air traffic as well as postal services and in July 1927 were transferred to the Department of Commerce, Bureau of Lighthouses, along with their facilities, now known as Airway Radio Stations. The stations were transferred in August 1938 to the Civil Aeronautics Authority, and they became Airway Communication Stations. Finally, these facilities became Flight Service Stations (FSSs) with the establishment of the Federal Aviation Administration (FAA) in 1958. The U.S. Weather Bureau became increasingly responsible for the collection and distribution of aviation weather information and forecasts. Pilots obtained briefings from the Weather Bureau and filed flight plans and acquired aeronautical information from the FSSs.

Because of the increase in air commerce and other factors, in 1961 the Weather Bureau began the certification of FSS personnel as pilot weather briefers. The FAA and the Weather Bureau signed a mutual

Memorandum of Agreement in 1965 delegating responsibility for pilot weather briefing to the FAA. FSS controllers had little in the way of guidelines during this period regarding the structure of weather briefings; they basically read weather reports and forecasts verbatim as requested by the pilot.

A field FSS pilot briefing deficiency analysis group began a special evaluation of pilot briefing services in 1975. The group cited deficiencies in the use of a standardized briefing format. (The standardized format had been taught at the FAA Academy for some time; however, it had yet to be incorporated in the FSS Handbook.) Other areas identified were the reading of weather reports and forecasts verbatim as opposed to interpreting, translating, and summarizing data. A poor level of proficiency in reading, understanding, and employing facsimile charts was noted. Briefers failed to obtain sufficient background information to tailor briefings to the types of flights planned.

From this study came the FAA's emphasis on an extremely rigid briefing format and an ambitious refresher training program. Unfortunately, the FAA did little to inform pilots or other offices within the agency of this change in policy. This led to a good deal of friction between briefers and pilots.

Over the years, however, through mostly local efforts, pilots have become acquainted with the standard briefing format. The refresher training was to be a continuing program conducted at least every 5 years. However, with few exceptions, this program has been abandoned, presumably due to fiscal constraints.

The National Transportation Safety Board (NTSB) also conducted a special investigation into FSS weather briefing inadequacies. In 6 of 72 accidents involving fatalities, the NTSB determined that pertinent meteorologic information was not passed to the pilot during the weather briefing. Basically, these deficiencies consisted of failure to pass weather advisories and icing forecasts and downplaying forecasts of hazardous weather. The result has been what many pilots consider to be overkill in disseminating these advisories.

A major change occurred in 1983, when the extremely rigid format was relaxed somewhat. Three types of briefings emerged:

- Standard briefing

- Abbreviated briefing

- Outlook briefing

In addition, the requirements for inflight briefings also were specified.

Almost from the time the FAA took over pilot briefing responsibility in 1965, the Service A (weather) teletype system was obsolete. Since that time, proposal after proposal has been made to update weather distribution. Even by the early 1980s, most FSSs still used the 100-word-per-minute electromechanical teletype equipment. Briefers had to sift through mountains of paper to provide a briefing, with weather reports as much as $1^{1}/_{2}$ hours old by the time they were relayed and available.

The FAA approved what was called the *interim* Service A system in November 1978, which had been tested at the Chicago FSS. Subsequently, the Service B teletype system for the transmission of flight plans and other messages was incorporated. Referred to as the *Leased A and B System* (LABS), it is in use at nonautomated FSSs. LABS was designed to update FSS Service A until a complete computer system could be developed and installed. LABS eliminates the need for the briefer to sort and post SPECIs, PIREPs, NOTAMs, and most amended forecasts. These housekeeping chores, which took considerable time, have been eliminated. With this system, most weather reports are available within 5 to 15 minutes of observation.

Development of *Model 1*, a completely computerized system, began in 1982, and the system came on-line in 1985. Automated Flight Service Stations (AFSSs) use this equipment.

According to the FAA, "The primary benefit of the (FSS automation) program is improved productivity through automation of the controller's access to detailed briefing information and flight plan filing. To some extent the improved quality of pilot briefings reduces the need for multiple briefings as in the past." Model 1 is in the same evolutionary category as ARTCC flight data processing was in the early 1970s. It takes care of many of the data-processing functions, such as flight plan transmission and tracking.

From a weather briefing point of view, however, it presents the same information that was available from teletype and LABS. With Model 1, amendments are more timely, but this will not be directly obvious to pilots. Model 1 does not necessarily improve pilot briefing productivity; in fact, productivity in certain cases is reduced. Model 1 is truly "user hostile" and presents information in much the same way as direct user access terminals (DUATs) and other commercial briefing systems.

In the late 1990s with the loss of the Miami AFSS due to hurricane Andrew and the St. Louis AFSS due to floods, Model 1 equipment not only was becoming unsupportable but very little existed. Model 1's replacement will be the Operational and Supportability Implementation System (OASIS). OASIS will provide essentially the same data, but now with its own weather graphics system. OASIS, like Model 1, allegedly will increase the quality and quantity of FSS briefing services. Although, OASIS does provide some very important ergonomic advances, do not count on improved quality or quantity unless the FAA decides to do some serious FSS controller training. To date, the system continues to be plagued with problems and delays.

Federal Aviation Regulations require each pilot-in-command, before beginning a flight, to become familiar with all available information concerning that flight. This information must include "For a flight under IFR or a flight not in the vicinity of an airport, weather reports and forecasts...and any known traffic delays of which the pilot has been advised by ATC."

Additional regulations specify fuel and alternate airport requirements. The regulations do not, however, require that meteorologic and aeronautical information be obtained from the FAA.

The FAA's standard briefing is designed for a pilot's initial weather rundown prior to departure. Standard briefings are not normally provided when the departure time is beyond 6 hours, nor current weather beyond 2 hours. It is to the pilot's advantage to obtain a standard briefing or update the briefing as close to departure time as possible.

Before beginning a briefing, the FSS controller must obtain background information that is pertinent and not evident or already known. The amount of information varies with the training and experience of the briefer, weather conditions, and the pilot's request. Pilots can assist the briefer and reduce delays by volunteering the following information:

1. *The type of flight planned.* Always advise the briefer if the flight can only be conducted VFR or if an IFR flight is planned or can be conducted IFR. Normally, the briefer will assume that a pilot is planning VFR, unless stated otherwise. Student pilots always should state this fact to help the briefer provide a briefing tailored to the student's needs. Also, new or low-time pilots and pilots unfamiliar with an area will receive better service if they so advise the briefer. This alerts the briefer to proceed more slowly, with greater detail.

2. *The aircraft number or pilot's name.* This is evidence that a briefing was obtained, as well as an indicator of FSS activity. In the absence of an aircraft number, the pilot's name is sufficient. Most briefings are recorded and reviewed in case of incident or accident; it is in the pilot's interest to get "on the record" as having received a briefing.

3. *The aircraft type.* Low-, medium-, and high-altitude flights present different briefing problems. This information allows briefers to tailor the briefing to a pilot's specific needs. By knowing the aircraft type, the briefer frequently can estimate general performance characteristics such as altitude, range, and time en route.

4. *The departure airport.* Pilots must be specific; they know the airport, but the briefer usually does not. This is important with FSS consolidation, 800 phone numbers, and in metropolitan areas.

5. *The estimated time of departure.* The estimated time of departure is essential, even if general, such as morning or afternoon.

6. *The proposed altitude or altitude range.* This information is needed to provide winds and temperatures aloft forecasts. If an altitude range is specified, e.g., 8000 to 12,000 ft, the briefer can provide the most efficient altitude for direction of flight.

7. *The route of flight.* The briefer will assume that a pilot is planning a direct flight unless otherwise informed. If an indirect route is planned, the pilot must provide the exact route or preferred route and any planned stops. This will assist the briefer in providing weather for the planned route.

8. *The destination airport.* Again, pilots must be specific. If not, a pilot may not receive all available weather and NOTAM information.

9. *Estimated time en route.* Many briefers can estimate time en route based on aircraft type. This information is needed to provide en route and destination forecasts. Total time en route is essential when stops or anything other than a direct flight is planned; for IFR flights, the estimated time of arrival is required to determine alternate requirements.

10. *Alternate airport.* If you already have an alternate in mind, provide it at this time. FSS equipment will automatically display alternate airport current weather, forecast, and NOTAMs to the briefer.

This may seem like a lot of information, but it really isn't. The briefer must obtain this information before or during the briefing. Providing background information will allow briefers to do a better job, which is to provide the pilot with a clear, concise, well-organized briefing tailored to a pilot's specific needs.

All right, the background information has been provided. What can a pilot expect in return? The briefer is required, using all available weather and aeronautical information, to provide a briefing in the following order. Pilots should be as familiar with this format as the mnemonic CIGAR (Controls, Instruments, Gas, Attitude, Runup) or the IFR clearance format.

1. *Adverse conditions.* Any information, aeronautical or meteorologic, that might influence the pilot to cancel, alter, or postpone the flight will be provided at this time. Items will consist of weather advisories, major NAVAID outages, runway or airport closures, or any other hazardous conditions.

The adverse conditions provided should only be those pertinent to the intended flight. This is one reason why the pilot must provide the briefer with accurate and specific background information. The briefer should then only furnish those conditions which affect the flight. There is, unfortunately, some paranoia among briefers that causes some of them to provide anything within 200 mi of the flight, whether it is applicable or not.

2. *VFR flight is not recommended* (VNR). Undoubtedly, the VNR statement is the most controversial element of the briefing; nevertheless, the FAA requires the briefer to "Include this recommendation when VFR flight is proposed and sky conditions or visibilities are present or forecast, surface or aloft, that (in the judgment of the specialist) would make flight under visual flight rules doubtful."

3. *Synopsis.* The synopsis is extracted and summarized from FA and TWEB route synopses, weather advisories, and surface and upper-level weather charts. This element may be combined with adverse conditions and the VNR statement, in any order, when it would help to more clearly describe conditions.

These three elements should provide pilots with the "big picture," part of the "complete picture." The synopsis should indicate the reason for any adverse conditions and tie in with current and forecast weather. During this portion of the briefing, pilots should pay particular attention for clues of icing, even if a weather advisory is not in effect. For example, if light icing is forecast, it may be overlooked.

4. *Current conditions.* Current weather will be summarized: point of departure, en route, and destination. Relevant PIREPs and weather radar reports will be included. Weather reports will not normally be read verbatim and may be omitted if the proposed departure time is beyond 2 hours, unless specifically requested by the pilot. One

should look for cloud bases en route and PIREPs of icing, freezing level, and tops. Forecast surface temperatures are not available at this time but may be in the future. However, one can extrapolate surface temperatures from current data.

5. *En route forecast.* The en route forecast will be summarized in a logical order (climbout, en route, and destination) from appropriate forecasts (FAs, TWEBs, weather advisories, and prog charts). The briefer will interpret, translate, and summarize expected conditions along the route. Specifically, pilots should look for forecast bases and tops.

6. *Destination forecast.* Using the TAF, where available, or appropriate portions of the FA or TWEB forecast, the briefer will provide a destination forecast, along with significant changes from 1 hour before until 1 hour after estimated time of arrival (ETA). Pilots should be especially interested in wind, visibility, weather, type of precipitation, and cloud bases.

7. *Winds aloft forecast.* The briefer will summarize forecast winds aloft for the proposed route. Normally, temperatures are provided only on request. Request temperatures aloft. A pilot would want to know if he or she is going to be below, at, or above the freezing level. Temperature at the flight planned altitude is an indicator of icing severity, as well as aircraft performance.

Remember that airframe ice will reduce airspeed. This could be a significant factor for flights at the maximum range of the aircraft. There was a Korean C-130 at Oakland for several weeks trying to get to Hawaii. It was winter, weight and fuel were at the outside of the performance envelope, and the aircraft could not accept any ice, even though it had ice protection equipment.

8. *Notices to Airmen (NOTAMs).* The briefer will review and provide applicable NOTAMs for the proposed flight that are on hand and not already carried in the *Notices to Airmen* publication. Pilots should pay particular attention to landing area conditions. One might want to check surface conditions for en route airports as possible alternates. This will require a specific request.

9. *Other services and items provided on request.* At this point in the briefing, briefers normally will inform pilots of the availability of flight plan, traffic advisory, and Flight Watch services and request pilot

reports. On request, the controller will provide information on military training route (MTR) and military operation area (MOA) activity, review the *Notices to Airmen* publication, check Loran or GPS NOTAMs, and provide other information requested.

It is not necessary to copy all the information provided because much is supplementary and provides a background for other portions of the briefing. Pertinent information should be noted, and it is often advantageous to copy these data. Many forms are available. It is often helpful to have a map containing weather advisory plotting points, as shown in Fig. 5-2.

Locations on the weather advisory plotting chart were changed on March 1, 1999. Figure 5-2 reflects these changes. However, the weather products used in this book preceded the change. Therefore, difference in the example weather products and Fig. 5-2 should be expected.

Briefers provide abbreviated briefings when a pilot requests specific data or information to update a previous briefing or supplement an FAA mass dissemination system (Transcribed Weather Broadcast, Telephone Information Briefing Service, or Pilot's Automatic Telephone Weather Answering Service).

When all that is required is specific information, a pilot should state this fact and request an abbreviated briefing. Because the briefer normally must make a request for each individual item, it is extremely helpful to request all items at the beginning of the briefing, thus reducing delays. The briefer will then provide the information requested. When using this procedure, the responsibility for obtaining all necessary and available information rests with the pilot, not the briefer. Pilots must realize that the briefer is still required to offer adverse conditions. Pilots sometimes become irritated when the briefer mentions weather advisories; however, this is an FSS Handbook requirement.

WEATHER ADVISORY PLOTTING CHART

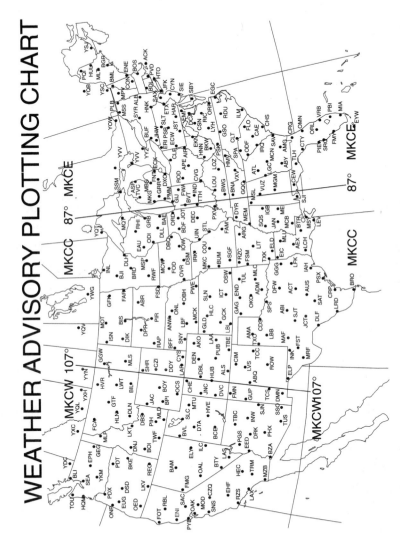

Fig. 5-2. To visualize weather-advisory-affected areas, it's often helpful to plot their location.

Pilots requesting an update to a previous briefing must provide the time the briefing was received and necessary background information. The briefer will then, to the extent possible, limit the briefing to appreciable changes. An alarming number of pilots, when asked the time of their previous briefing, respond, "I got the weather last night." Needless to say, this practice does not comply with regulations. These individuals should be requesting a standard briefing.

When requesting supplemental information for an FAA mass dissemination system, again, the briefer must have enough background information and the time the recording was obtained. The extent of the briefing will depend on the type of recording and time received.

With a proposed departure time beyond 6 hours, an outlook briefing normally will be provided. The briefing will contain available information applicable to the proposed flight. The detail will depend on the proposed time of departure. The farther in the future, the less specific is the briefing. At a minimum, the outlook will consist of a synopsis and route/destination forecast.

Although discouraged, unless unavoidable, briefings once airborne will be conducted in accordance with a standard, abbreviated, or outlook briefing as requested by the pilot. As with any briefing, sufficient background information must be made available to the briefer.

Briefings can be obtained in person, over the telephone, or by radio. The preferred methods are to obtain a weather briefing in person or by phone. Initial briefings by radio are discouraged, except where there is no other means. The reasons are simple. The cabin of an aircraft plunging into the wild gray yonder is no place to plan a flight. Attention must be diverted from flying the aircraft to the briefing. Especially with marginal weather, certain pilots have a tendency to push on, regardless of conditions, not to mention the fact that such a practice unnecessarily ties up already congested radio frequencies.

The weather briefing is a cooperative effort between the pilot and FSS controller. Preliminary planning should be complete, including a general idea of route, terrain, minimum altitudes, and possible alternates. Where available, pilots should obtain preliminary weather from one of the recorded services. From the broadcast, one should determine the type of briefing required—standard, abbreviated, or outlook.

During the briefing, try not to interrupt, unless the briefer is going too fast. Often pilots interrupt with a question that was just about to be answered. This can cause the briefer to lose his or her train of thought, resulting in an inadvertent omission of information.

Briefers make mistakes, and many are not pilots. At the end of the briefing, do not hesitate to ask for clarification or additional information on any point you do not understand completely. If conditions are right for turbulence and icing and these phenomena were not mentioned, ask the briefer to verify that there are no weather advisories. Remember that forecasts for light to locally moderate icing do not warrant an advisory, nor locally severe icing a SIGMET.

With this as a background and FSS staffing being further reduced, the question becomes how can a pilot best use the services available? The FAA authorized two companies in February 1990 to provide direct user access terminal (DUAT) service to pilots within the contiguous United States. This computerized system, available at airports and through personal computers, allows direct access to weather briefing and flight plan services. DUATs and virtually all other commercial services use National Weather Service (NWS) products, which contradicts the misconception that computer briefings somehow provide different products from those available through an FSS.

When using these services, it is essential for pilots to know what information is available. Pilots using a commercial system must check with the vendor to determine how the system handles aviation products. Certain advisories, e.g., CWAs, may not be available on some systems; none provide local NOTAMs. Know your service, and check with an FSS for any additional information required or to clarify anything you do not understand—remember the disclaimer.

The advantages of computer briefings are the relatively prompt access and the ability to retain a personal copy. With these advantages come the responsibility to decode, translate, interpret, and apply information to a flight. The pilot will have to sift through the mountains of written data, formally reserved for the FSS controller, to determine if a particular flight is feasible under existing and forecast conditions and aircraft/pilot capability.

The sheer amount of information may be overwhelming, especially for long-distance flights. A pilot may have to study several pages for a single sentence that applies. Finally, if there is a problem with one of these services, you must contact the vendor.

As so elegantly stated by John Hyde, an ex-Army aviator, Kit Fox owner, and Oakland FSS controller, when obtaining a briefing from an FSS or other source, keep in mind that they are in "sales not production." In other words, do not blame the messenger for the message.

Preparation

Some preparations for a flight with potential icing are the same as for any other flight; others apply specifically to the icing threat. This section focuses on preparations relating specifically to icing.

In order to apply a weather briefing to a flight, one must have done one's homework. What is the terrain like along the route? What are the minimum altitudes? Are there suitable alternates? What if plan A does not pan out? Therefore, it is incumbent on the pilot—for every briefing, but especially if there is possible icing—to study the terrain, routes, and possible alternates for the proposed flight. For example, recall my experience with Los Angeles Center. I had flight-planned the flight below the freezing level, but because of traffic, I had to climb into icing conditions. The objective is to have an out. If there are no "outs," the flight is a definite no-go.

In order to apply a weather briefing to a flight, a pilot must consider the aircraft he or she is planning to fly. Is the aircraft ready for cold weather and potential icing operations? If not, it's a no-go. How about the pilot and passengers? Human factors are often overlooked and take on additional significance during cold weather and icing conditions.

Evaluating the Weather

Now we can apply our knowledge of the physics of icing, icing reports and forecasts, and the weather briefing to flight situations. With icing, the goal is to minimize exposure.

Pilots planning flights below the freezing level normally cannot expect to receive an icing advisory during an FSS preflight briefing, since icing will not affect their proposed flight. Some briefers fail to understand and consider this and issue the advisory even though it is not a factor. This practice undermines the credibility of both the forecast and the briefing. Pilots planning flights and briefed for low altitudes should keep this point in mind, in the event they should elect, or be instructed by air traffic control (ATC), to climb to a higher altitude. This is another reason to request temperatures for the planned cruising altitude. From this information the pilot can determine how close the temperature is to the freezing level. By comparing forecast temperature with observed temperature, the pilot can get a sense of forecast accuracy. A large discrepancy may indicate a "busted" forecast. Pilots might well consider the advisability of accepting clearance without additional information on icing and freezing levels. This point also applies to rerouting. Should the pilot or ATC reroute the aircraft, advisories that were not pertinent during the briefing may now apply.

Previously, escape routes out of icing were mentioned. From the preflight briefing, prospective "outs" fall into the following categories:

- Plan a course to minimize exposure to icing.

- Climb above cloud tops.

- Climb to cold air.

- Descend below the freezing level.

- Avoid clouds and precipitation at below-freezing temperatures.

- Return to an area of ice-free air.

Based on the weather briefing and a "complete picture" of conditions, plan the route of flight to avoid potential areas of icing, penetrate frontal areas at as close to a 90° angle as possible to minimize icing, circumnavigate areas of potential icing, land short and wait until the icing hazard no longer exists, or delay the flight until the icing hazard has passed.

Should icing be encountered, a first choice may be to climb above the clouds into clear air. This assumes that the pilot knows where the tops are and the aircraft has the capability to climb through an icing layer to this altitude. The pilot should climb to air that is too cold for icing. This, again, assumes the previous caveats. Descend below the freezing level. This requires a knowledge of the freezing level, terrain, and minimum altitudes. One can avoid an icing hazard by remaining clear of clouds and precipitation when flying above the freezing level. ATC always seems to have a way of destroying "the best laid plans." Like the Boy Scouts, one should "be prepared." Finally, if necessary, one can turn around, presumably having come from an ice-free area.

Again, minimize exposure. Should ice be encountered, immediately notify ATC and initiate a plan of action. The point is to do something.

These points were supremely illustrated in two scenarios in Chapter 3. Recall the Baron pilot who, without ice protection

> **CASE STUDY**
> Among the many pilots with whom I have had the pleasure of
> serving at the Oakland FSS was a particular local air taxi pilot.
> He flew routinely from Oakland, through the Sacramento Valley,
> to northern California. He provided all the necessary
> background information and always requested a standard
> briefing. Since it is not part of a standard briefing, he would
> always ask for the closest area of clear conditions at the
> conclusion of the briefing. What an excellent idea! This was one
> of the most prepared pilots I have known. In the event that he
> had engine, navigation, or electrical problems, he knew the
> closest location of clear weather. All pilots should add this
> technique to their personnel briefing requirements.

equipment, departed Reno, Nevada, for southern California. This pilot
elected to fly a direct course along the crest of the Sierra Nevada
Mountains, the route where the most intense icing and turbulence
could be expected.

The pilot had no way out because the MEA was the aircraft service
ceiling. The terrain was well above the freezing level. The pilot failed
to reverse course at the first sign of ice. What other options were
available? The pilot could have crossed the mountains near
Sacramento, minimizing exposure to ice and at a 90° angle to the
weather system. Once over the Sierra Mountains, it was all downhill.
The pilot could have flown toward Las Vegas, where the weather was
considerably better, or simply waited for better weather conditions.
When the airplane became ice covered, the pilot had no option but to
ride it to the crash site.

Remember the Bonanza pilot who departed the San Francisco Bay
Area on a flight to Los Angeles. This pilot elected to fly at 11,000 ft,
well into the icing layer. The pilot's last words were, "I've iced up and
stalled." The crash occurred where the elevation was near sea level.
Minimum altitudes were well below the freezing level. The pilot
simply did nothing until aircraft control was lost.

An advantage of turbocharged and pressurized aircraft is the ability to fly high, above the weather. Icing normally is not a significant factor in such flight levels, except around convective activity or in the summer when temperatures can range between 0 and $-15°C$ at these altitudes. Just because it is summer does not necessarily mean that there is no icing potential.

CASE STUDY

A pilot was flying a Mooney from Little Rock, Arkansas, to Charleston, West Virginia. The pilot's ultimate destination was Massachusetts. The briefing included weather advisories for turbulence and light to moderate rime icing below 12,000 ft in clouds and precipitation. Scattered to broken clouds were forecast at 2000 to 3000 ft. For Charleston, *26009G21KT 7SM - SN BKN027 OVC040 01/M04....* The briefer editorialized about conditions "not looking too bad." The pilot proceeded on top for the western portion of the flight.

The Mooney was on frequency when the center received a PIREP from a pilot departing Charleston reporting moderate rime ice. As the flight approached Charleston, the ATIS reported 26010KT 7SM -SG OVC013 01/M01 A3000. Charleston approach control received numerous reports of light to moderate mixed and rime icing.

The Mooney's pilot was cleared to descend and received vectors for the approach. On descent, the pilot lost control. The airplane was reported to have rocked from side to side and crashed.

The pilot received the essential information—icing below 12,000 ft. The temperature at Charleston was right at the freezing point, with light snow reported. Any ice would be carried all the way to ground. The pilot was flying from good weather to bad. Upslope, caused by the rising terrain of the Appalachian Mountains, would enhance any icing present. PIREPs confirmed the forecast. The pilot had the opportunity to land short or reverse course but exercised neither of these options. If the pilot had landed short, there was a good possibility of overflying the Appalachians. The pilot could then land closer to the East Coast, in an area not affected by upslope and with higher ground temperatures. Although we will never know, it appears that this pilot was locked into one, poorly conceived plan.

Some pilots say never say never. Here is the exception that proves the rule. Never let the briefer make the go-no-go decision. The briefer is a resource, and some are better than others. This applies equally to optimistic and pessimistic briefers.

For example, let's examine the "VFR flight is not recommended" statement. It leaves considerable leeway for the briefer; some use this statement more than others. The inclusion of this statement should not necessarily be interpreted as an automatic cancellation, nor its absence as a go-for-it day. Notice that VNR applies to sky condition and visibility only. Few understand the provisions of special VFR. Hazardous phenomena, such as turbulence, icing, winds, and thunderstorms, of themselves, do not warrant the issuance of this statement. It is important to remember that this is a recommendation. Why, then, use such a statement? It's simple, every year pilots insist on killing themselves and their passengers at an alarming and relatively constant rate by flying into weather where they have no business.

This statement was instituted in 1974, presumably because the last person a pilot would talk to usually was the briefer. A logical, although alarming, result of this statement is the increasing number of pilots who, in the absence of VNR, ask, "Is VFR recommended?" So far, the answer remains that the decision as to whether the flight can be conducted safely rests solely with the pilot.

It has been my experience that VFR flight is possible between 50 and 60 percent of the time that this statement has been issued for flights that I planned VFR. Remember that this is based on my training and experience. This does not mean that the statement should be ignored, but pilots must take a careful look at the "complete picture." This issue will come up again in later discussions of risk assessment and management.

Surface wind reports and forecasts are an important part of weather evaluation for flight to and from areas with surface snow, ice, or slush.

It is essential to calculate crosswind or tailwind components. Significant crosswinds or tailwinds may result in a no-go decision.

Crosswind or tailwind components can be determined using a flight computer or a graph, as illustrated in Fig. 5-3. This figure allows pilots to determine crosswind, headwind, and tailwind components. The vertical and horizontal axes represent wind speed in knots; the arc represents crosswind angle in degrees. Three elements are required: runway orientation (i.e., runway number), magnetic wind direction, and wind speed.

For example, suppose that a pilot is planning a departure from runway 25 (runway magnetic orientation 250°). The ATIS reports surface wind from 280° at 17 knots (*28017KT*). If the wind was obtained from a METAR, the pilot would have to calculate the difference between runway orientation and true wind direction. (It does not matter if it is a left or a right crosswind, what is needed is the difference between runway heading and wind direction.) In this case, the difference is 30° (280 − 250 = 30). The pilot enters the crosswind component chart on the 30° radial line and proceeds toward the origin of the graph—toward zero. Progress stops at the point where the 30° radial line intercepts the wind speed arc of 17 knots. Moving vertically down to the horizontal axis, the pilot reads 9 knots. The crosswind component for this runway and wind is 9 knots. By moving horizontally, to the vertical axis, the headwind component is approximately 14 knots. Should the angular difference between runway and wind be more than 90° but less than 180°, the pilot would have a crosswind and tailwind component.

Most airplanes have a published "maximum demonstrated crosswind component." This is not an absolute limit but should be considered along with the pilot's training and experience. All airplane flight manuals (AFMs) contain takeoff and landing performance charts for various headwind, runway, and temperature conditions. Some AFMs also contain performance charts for takeoff with a tailwind.

CROSSWIND COMPONENT CHART

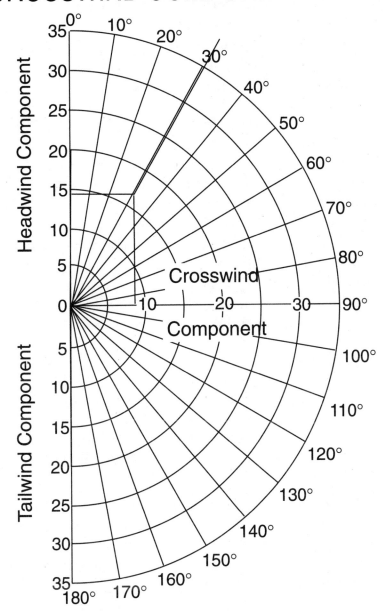

Fig. 5-3. Crosswind or tailwind components take on greater significance with a snow-, ice-, or slush-covered runway.

Aircraft and Engine Considerations

If you are based in a warm climate, you may not be familiar with the aircraft manufacturer's recommendations for winterizing your aircraft. Most mechanical equipment, including aircraft and components, have specific design temperatures. Know and follow the manufacturer's recommendation.

Some manufacturers recommend engine and oil covers and baffles. When baffles are installed, a cylinder head temperature gauge is recommended, particularly if wide temperature differences are expected. Engine oil is extremely important in low temperatures. Be sure the proper weight oil is used in low temperatures. Pay particular attention to the crankcase breather in cold weather. A number of engine failures have resulted from frozen crankcase breather lines that caused pressure to build up, sometimes blowing off the oil fill cap or rupturing a case seal, causing loss of oil. Water vapor, a by-product of combustion, can condense and freeze. Special care is recommended during the preflight check to ensure that the breather system is free of ice.

CASE STUDY
The investigation of a Cessna 152 accident revealed that the crankcase breather line was plugged with ice and the oil had been forced out through the engine's nose seal. The probable cause was determined to be the pilot's inadequate preflight preparation that failed to detect the ice-clogged breather.

An important precaution for cold weather operations is an inspection of all hose lines, flexible tubing, and seals for deterioration. After replacement of all doubtful components, be certain that all clamps and fittings are torqued properly to the manufacturer's specifications for cold weather.

Many aircraft are equipped with cabin heater shrouds that enclose the muffler or portions of the exhaust system. It is imperative that a thorough inspection of the heating system be made to eliminate the possibility of carbon monoxide entering the cabin.

Wet-cell batteries are of special concern during cold weather. They should be kept fully charged or removed from the aircraft when it is parked outside to prevent loss of power caused by cold temperatures and possible freezing.

A dead battery can be very annoying, especially in cold weather. With a dead battery, a pilot may have one of three options: charge the battery, use an external power source, or handprop the airplane. Assuming that there is a maintenance shop on the field, a pilot may wish to change the battery. Unfortunately, this takes considerable time. Many airport operators have auxiliary power units (APUs). If the aircraft is equipped with an auxiliary power receptacle, this may be the easiest and quickest solution. Be sure to follow the manufacturer's recommendations. If the airplane has a generator or alternator with some power left in battery, handpropping is another alternative.

FACT

Each year accident investigations have revealed that carbon monoxide has been a probable cause in accidents that have occurred in cold weather operations.

WARNING

Handpropping is an extremely dangerous procedure. Handprop an airplane only when it is absolutely necessary and only after taking proper precautions. Never handprop unless a qualified person, thoroughly familiar with the operation of all controls, is seated in the airplane with the brakes set. Leave the wheels chocked and at least the tail tie-down secured. The ground should be firm and free of debris. Loose gravel or a slippery surface (ice and snow) may cause the person to slip or fall into the propeller. If you think that you may need to handprop your airplane, obtain instruction from a qualified flight instructor or mechanic.

Remove all ice, snow, and frost. Recall my experience at South Lake Tahoe with the ice-covered Cessna 172. Freshly fallen snow can be brushed off; ice and frost must be scraped off or melted. Day-old snow tends to freeze to the aircraft surfaces, requiring deicing prior to flight. Alcohol or one of the ice-removal compounds may be used. Exercise caution if the aircraft is taken from a heated hanger and allowed to sit outside for an extended length of time when it is snowing. Falling snow may melt on contact with the warm aircraft surface and then refreeze. It may look like freshly fallen snow, but it will not blow off during the takeoff roll.

Aircraft put into heated hangers to melt frost, snow, and ice must be dried if temperatures at the surface or aloft are below freezing. Any water left on the aircraft will refreeze once the aircraft leaves the hanger.

Snow is heavy. Wings are designed to support an airplane from beneath and usually are not designed to support a heavy load from above. If snow accumulates on the wing, it can cause damage to the structure. Carefully inspect wing and tail surfaces for damage if they have been subjected to a large amount of snow.

Some pilots use warm water to melt snow and ice. The snow melts, but there is a possibility that the water can refreeze, leaving a glaze of clear ice on the surface. Dripping water also can freeze on control actuators, wheels, and brakes. Should one wheel ice up, you literally could end up going in circles!

Sweep snow off the aircraft with a broom, and then use a deicing compound to flush away any remaining ice. On fabric-covered airplanes, avoid using anything that could crack or bend the fabric. If available, use wing and tail covers.

If an aircraft is parked in an area of blowing snow, special attention should be given to openings in the aircraft where snow can enter, freeze solid, and obstruct operations. These openings must be free of snow and ice before flight. Some of these areas are

- Pitot tubes/static ports
- Heater/air vent intakes
- Fuel tank vents
- Carburetor intakes

CASE STUDY

A witness observed a Piper Comanche during its initial climb after takeoff, with landing gear extended. A flight instructor, flying above and behind, observed the airplane pitch up and enter a two-turn left spin. The airplane crashed in a steep descent. Before the flight, the pilot requested that a line attendant brush snow off the airplane. After the snow was removed, light icing remained on the inboard wing and left inboard horizontal stabilizer. The airplane was loaded about 125 lb over its maximum gross weight.

The probable cause was the pilot's inadequate preflight preparation. The inadequate removal of airframe ice resulted in a stall/spin and collision with terrain. Contributing factors were the pilot's failure to raise the landing gear during the climb after takeoff and loading the airplane over its allowable gross weight.

- Antitorque and elevator controls
- Main wheel and tail wheel wells
- Openings around control surfaces
- Propeller spinners

It is human nature in cold temperatures to hurry the preflight inspection of aircraft and equipment. However, this is the time to pay particular attention to this task. Watch for water or ice in fuel tanks, fuel lines, and sumps.

Our local crop duster pilot at Lovelock, Nevada, parked his Grumman Agcats during the winter. Periodically, he would come out to the airport and run the engines. On one occasion, it was freezing with a 20-knot wind. This puts the wind chill factor at about $-15°C$ (Fig. 5-4). He cranked and cranked and cranked the engine. It finally started, but would barely run. I found it very humorous to see him climbing out on the wing, with the engine sputtering, draining the fuel sumps—I doubt if he saw any humor in the incident.

WIND CHILL FACTOR

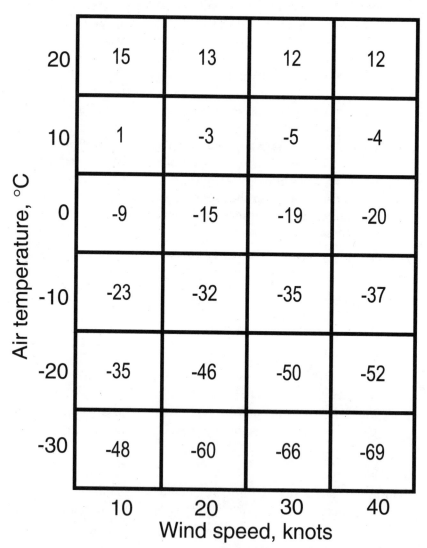

Fig. 5-4. Wind chill factor is the apparent lowering of the air temperature sensed by our body, caused by the wind.

Low temperatures change the viscosity of oil, batteries can lose a high percentage of their effectiveness, and instruments can stick. Preheating the engines and cabin before starting may be advisable. Extreme caution must be used in the preheat process to avoid fire.

Northern Nevada can be extremely cold in the winter. It was very cold that morning in Lovelock. There was a Cessna Skymaster on the field, and the owners wanted to fly over to Oakland for the football game. The airplane had not been flown for some time. They attempted to start the engine without success, and the battery was showing signs of stress. They decided to warm the rear engine by placing a flame heater at the rear of the aft engine, pointed into the cowling. Guess what? After a few minutes the rear engine caught fire! Fortunately, there was a fire extinguisher in the FSS.

The following precautions are recommended for preheating an aircraft:

- Preheat the aircraft by storing it in a heated hanger, if possible.
- Use only heaters in good condition, and do not fuel the heater while it is running.
- During the process, do not leave the aircraft unattended.
- Make sure a fire extinguisher is available.
- Do not place heat ducting so that it blows hot air directly on parts of the aircraft, such as upholstery, canvas engine covers, flexible fuel, oil, and hydraulic lines, or other items that may catch fire.

Follow the manufacturer's recommendations.

If the weather briefing indicates even a remote possibility of encountering ice, the pilot should accomplish several tasks: Ensure the pitot heat works with a very light touch during the preflight check. (Be careful, it gets hot!) Check the ice protection equipment for proper operation. Remember that a heated pitot is an anti-icing device, to be turned on before encountering ice; deicing equipment usually requires ice buildup before activation; improper operation actually can increase ice buildup and prevent its removal. Check alternate air or carburetor heat for an alternate source of air should the air filter ice over.

As a new instrument instructor, I took an instrument student on a flight from Van Nuys to Lancaster's Fox Field. The freezing level was forecast to be at 6000 ft. The minimum altitude for the route was 7000 ft. Cloud bases were at 6500 ft, well above terrain. Sure enough, we picked up trace to light rime icing. We had neglected to turn on the pitot heat, and a reverse cone of rime ice grew from the pitot tube. I, matter of factly, pointed this out to my student, and we turned on the heat, which immediately corrected the problem. On descent into Fox, the ice made a deafening nose as it broke off the tail surfaces. I do not recommend this procedure.

The pilot of a twin-engine airplane was en route when the airplane encountered icing conditions. When the pilot activated the deice boots, the right wing deice boot failed to function. As the airplane slowed for landing, the asymmetrical ice buildup caused instability that the pilot was unable to control. The airplane crashed on landing, seriously injuring the occupants. There was no mention of the pilot checking the boots prior to departure.

One of our local pilots at Fresno, California, had a similar occurrence in a Piper Aerostar. During the first storm of the season, on activating the deicing boots, one inflated, but the other did not. The pilot was visibly shaken by the experience. The bottom line: Check all aircraft equipment prior to departure—in accordance with the manufacturer's recommendations.

CASE STUDY

A pilot stated that during climbout on an IFR flight, he frequently checked the wings for ice accumulation, noting what appeared to be a "mere" trace of ice. At about 13,000 ft, with almost no climb at 125 knots indicated airspeed, the pilot pulled the yoke back a little to climb, and the aircraft began to shudder. The pilot then decided to divert, again checking the wings and seeing what he described as a trace of ice. He said he cycled the deice boots with no apparent change in wing appearance. While on approach, he again cycled the boots at between 3000 and 2000 ft. The pilot stated that about 20 to 30 ft above the surface, he started a slight roundout, but when he eased back on the yoke, the airplane shuddered and the nose did not rise. At that point he was ready to touch down, so he pulled back on the yoke while adding power; however, the airplane landed hard and was damaged. Later the surface deice system was tested. The boots did not hold pressure because of multiple holes and an internal shuttle valve leak. Functional tests revealed that the control valves did not direct proper pressure to the system and that the shuttle valves did not direct pressure properly to the boots. There was leakage to the vacuum side of the valves.

The probable cause, not surprisingly, was inoperative deicing equipment due to multiple wing boot holes and malfunctioning control/shuttle valves; subsequent buildup of airframe ice, and failure of the pilot to maintain adequate airspeed during the flare for landing.

Pilot and Passenger Considerations

Besides the aircraft, winter flights must consider the pilot and passengers. Is everyone dressed properly for the environment? In addition to personal comfort, is everyone psychologically and physiologically prepared to "commit aviation."

To begin, consider proper apparel. Dress for flying. This applies equally to pilots and passengers. If you can inspect the airplane and not get dirty, you have not done a thorough job! Wear slacks. Shoes should be flat-soled for safety and to ensure proper flight control operation. Avoid loose-fitting clothes or jewelry that could get caught on sharp edges around the airplane.

It can get mighty uncomfortable on cold and windy days, and normally, temperature decreases with altitude. *Wind chill factor* is the cooling effect of temperature and wind. Wind has the effect of lowering the apparent temperature sensed by the body. This is graphically depicted in Fig. 5-4. Recall the crop duster at Lovelock mentioned earlier, who was working in air at a temperature of about 0°C with a 20-knot wind. This resulted in a wind chill factor of −15°C.

Cabin heat normally is obtained by routing outside air through a muffler shroud that surrounds the engine exhaust stacks on most light, single-engine aircraft. This raises the temperature of the air by about 20°C. To illustrate, consider a trip from Reno to Lovelock, Nevada. At 7500 ft, the outside air temperature (OAT) was −18°C. The air entering the cabin was between 0 and 5°C. Bring a wind breaker or jacket when conditions warrant.

Are you flying over sparsely populated or mountainous terrain? Then you should carry waterproof jackets, long pants, boots, and gloves. Do you have proper survival gear for a forced landing? Your survival might depend on being properly equipped for an emergency landing, in below-freezing conditions, and being able to survive until rescued.

Years ago an airplane crashed in California's Sierra Nevada Mountains. This incident subsequently was made into the TV movie, *I Alone Survived.* The accident perfectly illustrates the hazards of flying over wilderness without proper clothing or survival equipment. There were two fatalities. The flight was from the San Francisco Bay Area to Death Valley, both relatively warm areas. The crash occurred on the crest of the Sierra Nevada Mountains at an elevation above 11,000 ft. The pilot and passengers had no survival gear or proper clothing. The sole survivor had to walk many miles out of the mountains.

Pilots in the southwest have to be especially aware of the hazards associated with a crash landing. We typically fly from warm, populated coastal areas and valleys to freezing, snow-covered wilderness areas. General aviation pilots are not the only ones seduced by our climate. The Navy has a major training facility at Lemoore, California. Navy pilots routinely fly to the gunnery and bombing ranges, across the Sierra Mountains to northern Nevada. Occasionally, they get lax and wear their flight suits over their skivvies. To keep this practice from getting out of hand, the Navy occasionally takes pilots by helicopter to terrain at the 8000-ft level for an overnighter, with only what they are wearing and their survival gear.

It is relatively inexpensive to put together a first aid/survival kit and carry a nonbreakable jug of water. This, along with proper clothing for the terrain to be flown, should be adequate. According to

Murphy's law, the only time you will ever need this equipment is when you did not bring it.

Everyday illnesses can seriously degrade pilot performance. Illness can produce distracting symptoms that impair judgment, memory, alertness, and the ability to make calculations. Even if symptoms appear to be under control with medication, the medication itself often impairs performance.

Did you know a visit to your dentist, with the accompanying pain reliever, may seriously impair your performance? Cough suppressants behave in the same way. The safest rule is not to fly while suffering from any illness. If you have questions about a particular malady, contact your aviation medical examiner for advice.

Minimum time between alcohol consumption and flying is contained in the regulations. However, as is often the case, minimum does not necessarily equate to safe. Research indicates that as little as 1 ounce of liquor, one bottle of beer, or 4 ounces of wine can impair flying skills. Alcohol consumed in these drinks is detectable in the blood and on the breath for at least 3 hours. Alcohol also renders pilots much more susceptible to disorientation and hypoxia.

Day-to-day living experiences also affect a pilot's flying ability and safety. How? Well, get up at "oh-dark-thirty," go to work, work a full day—trying to get out as much work as possible—drive to the airport, and begin the preflight inspection. Think about it. We just drove to and from work; freeway traffic—someone just cut you off; they are probably on their way to the airport to go flying—and to say the least, you are perturbed. Now, continue with the flight.

Fatigue—believe it or not—can be a serious problem. Fatigue can be described as either acute (short term—gone after a good night's sleep or perhaps a nap) or chronic (long term—those all-niters preparing for final exams or partying). This is just one of many everyday living occurrences that can cause fatigue. Fatigue is the tiredness felt after physical or mental strain, including muscular effort, immobility, heavy mental workload, strong emotional pressure, monotony, and lack of sleep. Fatigue can be minimized with proper rest and sleep, regular exercise, and proper nutrition (M&Ms, Coke, coffee, and donuts do not count).

Other health factors related to safety are stress and emotional well-being. Stress from daily activities, such as work or home management, typically is not relieved by flying. Stress and fatigue can be an extremely hazardous combination.

Emotions, upset by events such as a serious argument, death, separation or divorce, loss of a job, or financial problems, also affect a pilot's ability to fly safely. If you experience an emotionally upsetting event, you should not fly until you have given yourself adequate time to recover.

A pilot's involvement in all the health aspects of flying continues until the day he or she hangs up his or her certificate. Pilots are prohibited from flying with any known medical condition that does not meet the standards of their medical certification.

The FAA has developed a somewhat hokey but nonetheless useful physical and mental checklist. A pilot is safe to fly by not being impaired by

- Illness
- Medication
- Stress
- Alcohol
- Fatigue
- Emotion

I'M SAFE! (I told you it was hokey.)

If you're not at your peak, seriously consider canceling and rescheduling your flight. Please believe me, you can really mess up your day—in more ways than one—if you do not cancel.

Any discussion of fitness for flight would not be complete without a word about hazardous attitudes. Volumes have been written on the subject; you are encouraged to do more research into the matter, to know as much about it as possible. What is it? Research shows that most preventable accidents have one common factor, or as HAL9000 would put it, "human error." All too often pilots are their own worst enemies.

There are five identified attitudes that adversely affect a pilot's ability to make sound decisions. The first is called *macho.* This is an attitude in which an individual thinks that he or she must continually demonstrate that he or she is better than others. This usually results in unsafe actions. Although typically associated with men, women can be just as susceptible.

Closely associated with macho is *antiauthority*. These individuals believe that the rules do not apply to them. This attitude manifests itself as "Don't tell me what I can do!"

As experience grows through training and experience, pilots gain respect for aviation, the nature of flight, and their own mental attitude. Hopefully, the mental attitude that a pilot develops will be positive. But pilots also must be careful not to slip into the third of negative attitudes: *invulnerability.* Some people feel that bad things only happen to the others. I hope that everyone can see how a combination of macho, antiauthority, and invulnerability constitutes a road to disaster!

FIVE DANGEROUS ATTITUDES

- Macho
- Antiauthority
- Invulnerability
- Resignation
- Impulsiveness

The last two negative attitudes are *impulsiveness* and *resignation.* Impulsive individuals feel that they must do something, anything, immediately! A prime example is the failure of an engine on a multiengine airplane. Pilots can become so impulsive that they shut down the good engine or stall the airplane. It has happened to air carrier pilots, as well as to general aviation pilots. They do not take the time to evaluate the situation or consider options and risks before taking action. Finally, people who just give up demonstrate the attitude of resignation. It is their fate or bad luck. An example is the pilot who lost his attitude and heading indicators while flying in IFR conditions. He advised ATC, "I've lost the gyros, we're going in." Good thing the astronauts and mission control on *Apollo 13* were not afflicted with this attitude.

The first step in correcting a hazardous attitude is to recognize the behavior. Work with your instructor to identify any negative attitudes and change them.

Risk Assessment and Management

In the area of risk, it is often helpful to look at statistics. This is not the end of assessment and management but the beginning. In all icing accidents, almost half the pilots involved had more than 1000 hours of experience. Add to this that over one-third of the accidents were caused by pilots continuing flight into known icing. This indicates that individuals at risk in icing accidents have a good deal of experience. Using myself as an example, I gained quite a lot of training and experience, without being exposed to or trained for operation in an icing environment.

> ## CASE STUDY
>
> A Cessna 320 was cruising at 18,000 ft when radar and radio communication were lost. Radar data indicated that the airplane made a 180° turn before crashing in a river. Witnesses reported that they heard the airplane and then saw it spiraling through the clouds. Postaccident examination revealed no evidence of a preimpact anomaly. Weather data indicated a temperature of −9 to −10°C at 18,000 ft with the freezing level 9000 to 13,000 ft. AIRMET ZULU was in effect for the route, and there were numerous PIREPs of icing conditions in the vicinity of the accident. The pilot received an abbreviated preflight weather briefing, which included the AIRMET. According to the owner's manual, pilots of this make and model airplane should avoid icing conditions whenever possible. The airplane was not equipped with deicing equipment, and the pitot heat switch was found in the off position.
>
> The probable cause was the pilot's continued flight into known adverse weather conditions and loss of control. Known icing was a related factor. It is reasonable to conclude that loss of the airspeed indicator was a factor in the pilot's loss of airplane control.

Over one-quarter of the pilots involved in icing accidents did not survive. It appears that not only did the test come before the lesson, but the results were fatal. Most accidents occurred during late fall and winter. This should not be a surprise, since these seasons produce the most icing weather.

Over half the icing accidents were due to carburetor/induction system icing. This accounts for the fact that two-thirds occurred in VFR weather conditions. A number happened during flight instruction.

CASE STUDY

A flight instructor stated that he had given his student a simulated engine-out emergency. The student completed the emergency and had initiated a climb from the low approach. The engine failed to develop power. A 1700-ft-long airstrip was selected for the "real thing." The airplane went off the end of the runway and collided with the ground. After the accident, the engine started and operated normally. No mechanical problems were noted during the wreckage examination. A review of weather data disclosed that conditions were favorable for the formation of carburetor ice. The flight instructor reported using the carburetor heat but during a subsequent conversation said that the engine was not cleared for more than 2 minutes during the descent for the simulated emergency.

This case study has ominous implications. Our cadre of flight instructors have not been training properly. And therefore, they are unable to train students properly! All these accidents are preventable.

A smaller but still significant number of accidents occur on airport surfaces. These involve snow, ice, and slush. Although less that 10 percent fall into this category, like carburetor icing accidents, they are all preventable.

Accident prevention is part of the National Aeronautics and Space Administration's (NASA) commitment to aeronautics. To this end, NASA has developed scenarios of precursors to aviation accidents. A *precursor* is a factor that precedes and indicates or suggests that an incident or accident will occur.

In Fig. 5-5, each "wheel" represents one precursor. It may be physical incapacity, poor judgment, aircraft deficiency, failure of the ATC system, the weather, or other factors which of themselves would not create an incident or accident but when taken together lead to disaster.

NASA's ACCIDENT PRECURSOR SCENARIO

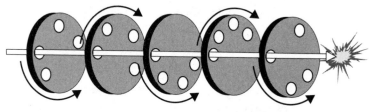

Alignment = Incident or Accident

Fig. 5-5. Precursors might be physical incapacity, poor judgment, aircraft deficiency, failure of the ATC system, or the weather.

CASE STUDY—JESSICA DUBROFF

Seven-year-old Jessica Dubroff accompanied her father (a passenger) and the pilot-in-command in an attempt to achieve a so-called transcontinental record involving 6660 mi of flying in 8 consecutive days. (I say so-called record because this was nothing more than a publicity stunt. It reminds me of telling friends that my son soloed at age 3 months. He was the sole occupant of the airplane as we pulled it over to the wash rack.) The first leg of the trip, about 8 hours of flying, had been completed the previous day, which began and ended with considerable media attention.

On the second day, the Dubroffs participated in media interviews, did a preflight check, and then loaded the airplane. The pilot-in-command received a weather briefing that included weather advisories for icing, turbulence, and IFR conditions because of a cold front moving through the area.

The airplane was taxied in rain for takeoff. While taxiing, the pilot acknowledged receiving information that the wind was from 280° at 20 knots, gusting to 30 knots. A departing Cessna 414 pilot reported moderate low-level windshear of ±15 knots. The airplane departed toward a nearby thunderstorm and began a gradual turn to an easterly heading.

Witnesses described the airplane's climb rate and speed as slow, and they observed the airplane enter a roll and descent that was consistent with a stall. Density altitude at the airport was 6670 ft. The airplane's gross weight was calculated to be 84 lb over the maximum limit at the time of impact.

> The probable cause was the pilot's improper decision to take off into deteriorating weather conditions. This included turbulence, gusty winds, an advancing thunderstorm, and possible carburetor and structural icing. The airplane was over gross weight. Density altitude was higher than that to which the pilot was accustomed. The result was a stall caused by failure of the pilot to maintain airspeed.

As in virtually all the preceding examples, most accidents can be attributed to a series of relatively insignificant factors that, when taken together, cause an accident. Let's review the Dubroff accident in this context.

The Dubroffs were on a tight schedule. Publicity events had been scheduled in advance. The original takeoff time was delayed to allow Jessica additional sleep. The pilot was fatigued from the previous day's flight and obtained little rest during the night. The weather was marginal at best. The pilot had to obtain a special VFR clearance for departure. Who was really flying the airplane? The pilot-in-command was seated in the right seat of the Cessna Cardinal. Now add high density altitude, an airplane over gross weight, and a mind-set that they must go.

The first precursor was the need to keep a time schedule—sometimes referred to as "get-home-itis." Precursor number two was pilot fatigue. The next precursor was a high-density-altitude takeoff with an airplane over gross weight. The fourth precursor was the weather, with its low ceilings and visibility, gusty winds, wind shear, turbulence, icing, and thunderstorms. (You could count each of these weather factors as an individual precursor.) A fifth precursor was the pilot's attempt, under these very adverse conditions, to try to maintain control of the airplane. We will never know exactly what happened, but airplane control was lost. The deck was certainly stacked against them.

Like most accidents, I think we can see how breaking any one individual link could have prevented this accident. The first link was the time schedule. A friend and excellent pilot has the philosophy that there is never a reason that you absolutely have to be anywhere. In the Dubroff accident, the pilot's mind-set appeared to be, "We're going no matter what."

The second link was fatigue. It was reported that Jessica had slept most of the first leg. As we have discussed, pilot fatigue is a significant factor in the deterioration of both mental and physical skills. This certainly may have clouded the pilot's go-no-go decision and failure to calculate gross weight and density altitude.

The weather was terrible. If the weather had been clear and calm, the pilot may have gotten away with fatigue, overloading the airplane, and lack of experience with high-density altitude.

Now add the pressure of flying from the right seat, with a novice student in the left, in less than VFR conditions. Even a slight, momentary distraction under these conditions can have serious consequences. It is reasonable to conclude that the pilot experienced sensory overload during climbout. All these factors together aligned the precursors, resulting in a fatal accident.

So how does one assess and manage risk? One applies *aeronautical decision making*—the ability to obtain all available, relevant information, evaluate alternate courses of actions, and then analyze and evaluate the risks and determine results. First, evaluate all the factors for a particular flight and decide if the risk is worth the mission. The goal is to prevent the precursors from aligning. This is easy for me to say. However, it can be extremely simple or extremely complex to achieve. There are three elements in risk assessment and management: planning, aircraft, and pilot. *Planning* is the "homework" part of the flight, which was discussed previously. Pilots study terrain, altitude requirements, and the environment. The environment includes the weather, the pilot's personal minimums, and alternatives. Then pilots evaluate the aircraft. Does it have the performance and equipment for the mission? If the answer is yes, the aircraft undergoes a preflight check to determine that it is airworthy. Assuming that the pilot is "fit for flight," the flight is ready to go. Simple, huh?

The decision can be as simple as my friend John looking at an afternoon flight in his Kit Fox in the traffic pattern or as complex as one of NASA's shuttle missions. To help, I have developed the *risk assessment and management decision tree* (Fig. 5-6).

Let's start with John's decision. *Planning:* Airport elevation 397 ft; runway 25L 2699 ft; pattern altitude 1400 ft; the environment—clear,

RISK ASSESSMENT AND MANAGEMENT DECISION TREE

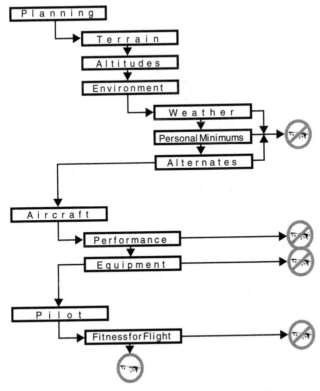

Fig. 5-6. Risk assessment and management can be as simple as a flight around the airport traffic pattern or as complex as a space shuttle mission.

cool, winds calm, alternate runway 25R. *Aircraft:* Performance of the Kit Fox OK; airplane equipped for flight in Class D airspace. *Pilot:* Fit for flight. Decision: Go!

Don't worry. I'm not going to evaluate a space shuttle mission. Instead, let's take an actual flight situation. I was flying from Oklahoma City to Palm Springs for the 1998 AOPA convention. You may recall that I made it as far as Santa Rosa, New Mexico, before the weather closed in. Hal Marx (USMC retired), the Santa Rosa Airport manager, fueled my airplane and gave me and my friends a lift into town, where we remained overnight. The next day was not any better, and we spent another night.

We had been trying to get to Albuquerque for 2 days without success. The following morning was not much better, but it was forecast to improve.

Planning: Santa Rosa, New Mexico, has a field elevation of 4782 ft. Along I-40, the high plateau of eastern New Mexico rises to over 7000 ft, with the pass through the Sandia Mountains at about the same elevation. Terrain is slightly lower to the north and south but still over 6000 ft. Because of the mountains, IFR MEAs vary from about 10,000 to 12,000 ft. Minimum altitudes would range from 6500 to 8500 ft VFR or 10,000 to 12,000 ft IFR. Even though I was flying a Cessna 172, I still had the option of going IFR or VFR.

Environment: Upslope due to rising terrain was, and continued to be, the culprit. MVFR to IFR ceilings, generally good visibility, high tops, freezing level at about 10,000 ft, conditions forecast to slowly improve during the day. When evaluating risk, flying toward or into improving weather is better than flying toward or into deteriorating conditions.

With my training and experience, I have different personal minimums depending on the environment. I also have confidence in my ability to make the decision to turn around. As John Hyde puts it, "Cowardice is the better part of valor" (undoubtedly an axiom he learned from his Army aviator days).

When one talks about personal minimums, there are a number of factors to consider. I have touched on some of them thus far:

When establishing personal minimums, first consider your level of training. Are you a student, private, commercial, instrument, or airline transport pilot? Our Air Force Aero Club had specific limits of wind, visibility, and ceiling for each certificate and rating.

PERSONAL MINIMUMS CONSIDERATIONS
- Training
- Experience
- Currency
- Aircraft
- Weather
- Time of day
- Physical condition
- Psychological condition

As level of experience increases, one may wish to consider different minimums. As a flight instructor, I tailored student minimums to their training and experience. For example, I had a student flying out of Lancaster's Fox Field, in the Mojave Desert of California. We trained in strong, gusty surface winds. When the student was proficient, I would increase the minimums. Some pilots obtain an instrument rating without ever having flown in the clouds. Do they have the experience to operate in actual instrument conditions? A prudent pilot would have another qualified, experienced pilot or flight instructor along until he or she became familiar with flight in clouds.

Currency with the type of operation is another personal minimum factor. Here again, legal does not necessarily mean safe. If you have been qualified recently to fly at night, you would certainly want to gain experience before tackling weather close to either VFR or IFR minimums during this time of day.

How familiar are you with the aircraft. If you have just checked out in a high-performance aircraft, especially without previous experience, are you ready to fly it in minimum weather? Probably not.

Low ceilings and visibilities, even when technically legal, are often an unacceptable risk. Depending on training and experience, low ceilings with good visibilities may be acceptable.

Time of day is another factor to consider. There is no question that flying at night introduces additional challenges and risk.

Physical and psychological condition were discussed earlier. If you are not fit, you should not go. Here is a good example of the application of personal minimums. One of my local flight standards operations inspectors had a flight in a Mooney 252 from Hayward to Ukiah, California. There was a low stratus layer over the San Francisco Bay. This individual had thousands of hours as a Navy P3 pilot. Even though this individual was qualified and current, he was not comfortable conducting this IFR operation. I volunteered to fly with him, and we had an uneventful flight.

IFR flight: High minimum altitudes, low freezing level, and I did not have approach charts for Albuquerque. The airplane would be at the limit of its performance envelope. The airplane was equipped for IFR operations, except not certified for flight in known icing. I would be at the MEA in probable icing conditions, unable to climb, over mountainous terrain. What alternates were available? None! Risk high. Decision: no-go.

VFR flight: Plan A—Climb to VFR on top and fly to Albuquerque and descend through broken clouds—forecast anyway. Plan B—Fly under the clouds and land at Albuquerque. Plan C—Fly south along the railroad to Albuquerque. (For some reason railroad engineers always select the lowest terrain.) Plan D—Return to Santa Rosa. Risk, yes, but plenty of options. For me this was a "go take a look" situation. Why? The area was sparsely populated, there was good visibility, and there was good weather at the departure airport. On the negative side, I was not familiar with the area; unfamiliarity has led many a pilot to disaster. It was daylight. A night flight, either IFR or VFR, under these conditions would have resulted in a no-go decision.

Airplane performance and equipment were go for the VFR plan. The pilot was fit for flight. Decision: go.

Risk assessment and management do not stop with a go decision. Pilots must reevaluate conditions throughout the operation, from preflight inspection to determining that a particular airport is suitable for landing. Should the airplane be unairworthy—this includes equipment—for the flight, the decision is no-go. If conditions at the destination (wind, weather, surface conditions, etc.) change, you may have to divert. If you do not have an alternate plan, risk is too high, resulting in a no-go decision.

With the preflight inspection complete and $4^{1}/_{2}$ hours of fuel, I departed and opened my VFR flight plan to Albuquerque. (A VFR flight plan, especially under these conditions, is part of risk management.) Ceilings were low, but visibility was excellent. It soon became apparent that plan A, over the clouds, was not going to work. This was confirmed through a conversation with Albuquerque radio advising that their weather had not improved. Plan A: No-go.

Plan B—Fly under the clouds. Approaching Clines Corners, terrain rises to about 7000 ft. The clouds went right down to the ground! When should a pilot say no and call it a day? I teach—or maybe it's preach—that the first time the thought occurs: "Should I really be here?" or "Maybe I should turn around," is a red flag to take positive action now. Do not push the weather, your aircraft, or yourself; turn around and wait it out. I initiated a 180° turn. I would have been flying from bad weather to worse weather. Risk too high. Decision: No-go.

At this point I had resigned myself to returning to Santa Rosa—plan D. However, my wife said, "What about plan C?" An increased risk accompanied plan C. There were only a couple of dirt strips with high elevations and short runways for alternates. The terrain was lower, ceilings were low, but visibility remained excellent. For navigation, I had the "iron compass" (railroad). I called Albuquerque radio to change our route and ETA. As is my practice, I made position reports and updated weather with flight service—another part of risk management. I always had the option of returning to Santa Rosa if the weather were to deteriorate.

Albuquerque did not improve, and I landed short at Alexander, New Mexico. With the weather now improving from the west, we continued on to Palm Springs.

The next chapter will continue to build on these principles of risk assessment and management.

ICING MYTH

Don't worry. The snow will blow off during the takeoff roll.

CHAPTER 6

Cockpit Strategies

This chapter explores cold weather operations and the effects of ice on flight operations from engine start to after landing. As in Chapter 5, flight planning strategies and cockpit strategies are often interrelated. Even though the focus of this chapter is primarily related to the taxi, takeoff, en route, and landing phases, occasionally the chapter refers back to topics that apply to flight planning. Otherwise, the chapter begins where Chapter 5 left off, with engine start and taxi. This is followed with a discussion of takeoff and landings. Since these two activities are closely linked, they are combined in one section.

Next the chapter explores the en route phase of flight. For our purposes, this begins after takeoff, with initial climbout from the airport, and ends with descent and approach. There continue to be pilots who blindly—again, pardon the pun—fly into icing conditions. Some recount how the initial weather briefing did not contain significant weather. Yet there is no mention of the pilot updating weather en route. Throughout the en route section I will discuss strategies for updating weather and applying risk assessment and management skills and decision making. The section concludes with strategies for descent and approach to landing.

> **IMPORTANT TIP**
> Continually updating the weather picture is the key to risk assessment and management during flight, especially in aircraft with limited or no ice protection equipment.

In the category of it isn't over 'til it's over, the next section discusses postflight considerations. In this section I'll present techniques and procedures for cold weather operations to prevent aircraft damage and prepare the ship for the next flight.

The final section takes a look at some actual flight scenarios. Like all the weather reports and forecasts contained in this book, all the scenarios were obtained during an actual weather event.

Winter can be one of the best flying seasons, with its cool temperatures and usually excellent flying weather between storms. However, like mountain flying in the summer, winter flying has its own unique problems and solutions. As well as aircraft icing, both VFR and IFR pilots may have to contend with a snow- or ice-covered runway. Additionally, airports may be closed for snow removal. Checking NOTAMs takes on even greater significance with airport closures, runway conditions, and breaking action reports.

Starting and Taxi

In moderately cold weather (0 to about −10°C), engines are sometimes started without preheat. Continental recommends preheating engines when temperatures drop below −1°C; Lycoming recommends preheat below −7°C. Oil is partially congealed, and turning the engine is difficult for the starter or by hand. Some manufacturers recommend pulling the propeller through by hand several times to clear the engine and limber up the cold, heavy oil before using the starter. They also may recommend priming the engine before pulling it through. (Use caution. Positively ensure that the magnetos are off. If possible, have a pilot at the controls. Check the surface around the propeller; it is easy to slip on a snow- or ice-covered surface. Recall the discussion of handpropping in the preceding chapter.)

Pulling the propeller through lessens the load on the battery. Without preheat, all available battery power is needed to start the engine. With a depleted battery, electrical reserve is reduced. Electrolyte in a discharged battery will freeze at a higher temperature than a fully charged battery. Therefore, keep the battery fully charged.

Prime the engine using the primer plunger. Wait about 60 seconds to allow the fuel to evaporate before cranking. Avoid pumping the throttle. Pumping the throttle to prime an aircraft engine is probably a holdover from automotive systems. Most aircraft carburetors incorporate an accelerator pump that squirts fuel up. Gravity causes the fuel to drip down and out of the carburetor, carburetor heat box, and cowl.

There is a tendency to overprime, which results in washed-down cylinder walls and possible scouring of the walls. This also results in poor compression and, consequently, harder starting. Sometimes aircraft fires occur from overpriming. This happens when the engine catches and the exhaust system contains raw fuel. Other fires are caused by backfires through the carburetor. It is good practice to have a fireguard handy during cold weather starts.

CASE STUDY

This situation occurred to me during my Lake Tahoe adventure. It was below freezing, and the engine was stubborn. Lack of experience at the time caused me to overprime. Raw fuel flowed from the carburetor, through the cowl, and onto the nose tire. During the next start attempt, the engine backfired. This started a small fire on the carburetor box and tire. My friends signaled, and I shut off the fuel and abandoned the airplane. I flagged down a passing airport truck equipped with a fire extinguisher. Naturally, the extinguisher was empty! I stuck my head down and blew out the flames. I am getting a little too old for this kind of behavior. I should have had a fireguard, with a tested fire extinguisher standing by. Then I should have kept cranking the engine to suck the flames into the engine as the fire guard extinguished the flames. It never ceases to amaze me how some people criticize inappropriate behavior, as if information is somehow transmitted by spores through the air. This is yet another example of where the test came before the lesson. If you are not familiar with cold weather operations, take of few lessons from a qualified instructor.

Unpreheated engines suffer from another cold start problem—iced-over spark plug electrodes. This happens when an engine only fires a few revolutions and then quits. There has been sufficient combustion to produce some water in the cylinders but insufficient combustion to heat the cylinders. This little bit of water condenses on the spark plug electrodes, freezes, and then shorts them out. When this occurs, the only remedy is heat. When no other heat source is available, the plugs can be removed from the engine and heated to the point where all moisture is removed.

Radios and other electrical equipment should not be turned on prior to engine start. All available power is needed to start the engine. Electrical equipment may be turned on after aircraft electrical power has stabilized and allowed to warm up. Cold weather is also hard on gyroscopic instruments. Cold weather causes excessive ware and requires time for such instruments to warm up. The solutions is to preheat the cabin or keep the airplane in a heated hanger.

If possible, select a location that is clear of debris and free of snow, ice, or slush. It may be a good idea to set the breaks and ensure that the aircraft remains fast. With the engine properly primed, a fire guard standing by, and the propeller area clear, start the engine.

After the engine starts, monitor oil pressure and cylinder head temperature; the use of carburetor heat may assist in fuel vaporization until the engine warms up. If oil pressure fails to come up to normal or drops after a few minutes, shut down the engine and check for broken oil lines, oil cooler leaks, or the possibility of congealed oil. During warmup it is possible to exceed cylinder head temperature, trying to bring up the oil temperature. Some engines require a minimum oil temperature and cylinder head temperature prior to runup and takeoff. Follow the manufacturer's recommendations.

Engines can quit during prolonged idling when insufficient heat is produced, fouling the spark plugs. Engines that quit under these circumstances are frequently found to have iced-over plugs.

Prior to taxi, monitor ATIS for surface condition NOTAMs. At uncontrolled fields, check with the airport manager or FSS for current conditions. If other aircraft are operating at the airport, request surface condition reports. Check wind direction and speed.

While taxiing, keep in mind that braking action on ice or snow is generally poor. If you cannot walk or drive on the surface, do not attempt to taxi. While driving to work at Lovelock, on a couple of inches of snow and ice, I attempted to brake for a turn. Nothing happened! I released the accelerator and brakes and let the snow decelerate the car. Fortunately, there was no traffic as I went several yards through the intersection. Braking action: Nil. In an airplane, your only option may be to shut down the engine and allow the airplane to come to a stop. Melting snow can refreeze from the previous day and form a sheet of ice. Ice from freezing rain can remain until removed or melted. Ice can produce a thin layer of water, lubricating the surface, making taxi or takeoff impossible.

Know your aircraft. Light aircraft can and do operate in dry snow up to 6 in deep. What depth of snow can your aircraft safely negotiate? This depends on a number of factors. Airplane propeller clearance is a significant factor. Mooneys have only about 6 in of clearance. There isn't much more with the Piper Arrow shown in Fig. 6-1. Wheel pants may look nice, but they allow only a few inches of clearance. Note in Fig. 6-2 that the Cessna 172 has very little clearance between the ramp and the wheel pants. Be especially cautious of snow banks and *berms*—a raised bank, usually frozen snow, along a taxiway or runway—and ditches while taxiing or taking the runway. Never attempt to taxi through a snow bank. Additional caution is required with low-wing airplanes. The center portion of taxiways and runways may be clear but may not allow wing clearance.

On icy taxiways and runways, even a light crosswind (5–10 knots) can cause directional control problems. If it is necessary to taxi downwind with either wheels or skis with a strong wind (10–20 knots), get help or do not go. When operating on skis, you have no

Fig. 6-1. When operating in snow, keep in mind that airplane propeller clearance may be only a few inches.

Fig. 6-2. Wheel pants only allow a few inches of clearance and may become clogged with snow and ice, causing brakes to lock.

brakes and no traction in a crosswind. On a hard-packed or icy surface, the aircraft will slide sideways in a crosswind, and direction control is minimal, particularly during taxiing, when control surfaces are ineffective. Large aircraft are not immune to these conditions. Several years ago a Boeing 747 slid off the runway at Anchorage, Alaska, due to icy conditions and a crosswind.

Aircraft, equipment, and personnel operating on a slick or icy surface may not have sufficient traction to start, stop, or even remain in place when encountering propeller wash—you know the stuff that keep the props so shiny—or jet blast. Maintaining directional control under these conditions is difficult at best. If at all possible, avoid taxiing behind other aircraft with running engines. How do you know the engines are running? For many years a recommended good operating practice has been to turn on aircraft anticollision lights just prior to engine start and while the engines are running. This procedure takes on greater emphasis during icy conditions. Remember that parked or holding aircraft, especially turbojets, require increased power to break away, maneuver, and taxi under icy surface conditions.

The absence of visible painted markings or obliterated signs can make maneuvering on the surface difficult. Unable to see visual aids, pilots may be hard pressed to judge direction and obstacle clearances. Added caution is required during these conditions. If you are not sure where to taxi, request assistance from ground control or UNICOM. In their absence, it may be necessary to shut done the airplane, off the active runway, and obtain assistance or have someone lead you to the ramp.

During taxi, avoid sharp turns and quick stops. Do not taxi through small snowdrifts or snow banks along the edge of the runway. They often hide areas of solid ice beneath the snow. Hitting an ice berm could result in serious propeller and engine damage. While operating on skis, avoid sharp turns; this may put excessive loads on the landing gear. Avoid areas of standing water or slush. Mud and slush thrown into wheel wells, wheel pants, and control surfaces can freeze, resulting in locked controls and frozen landing gear or brakes.

Takeoff and Landing

Takeoffs in cold weather offer some distinct advantages but also present special problems. Pilots must avoid overboost on super- or turbocharged engines. Density altitude often can be below sea level. Pilots need to exercised care when operating normally aspirated engines as well. Power output increases at about 1 percent for each 8°C below standard. At −40°C, the engine develops 10 percent more power, even though rpm and manifold pressure limits are not exceeded.

Consider the use of carburetor heat. Recall that in some cases it may be necessary to use heat to vaporize fuel. However, avoid the use of carburetor heat that raises the mixture temperature to freezing or just below. This may induce carburetor icing. A carburetor air temperature gauge is helpful. It may be best to use carburetor heat on takeoff in extreme cases of very cold weather. Follow the manufacturer's recommendations.

Takeoffs and landing on runways with standing water, snow, ice, or slush can be extra hazardous. These conditions impede aircraft acceleration. Although acceptable limits vary by aircraft, most jet aircraft flight manuals limit aircraft takeoff to $1/2$ in or less of standing water or slush and landing to 1 in or less on the runway. Light aircraft have no such testing or limitations, but these guidelines should be seriously considered.

Takeoffs should not be attempted in standing water or wet snow greater than $1/2$ in deep that covers an appreciable part of the runway. Additional ground roll will be required for takeoffs. Like high-density-altitude takeoffs, select a point to abort the takeoff with sufficient runway to stop if the airplane is not out of ground effect with a positive rate of climb. A soft field technique should be considered; always follow the manufacturer's recommendations, when published.

Hydroplaning results when a thin layer of water exists between the runway and the tire. Tires lose contact with the surface, and the airplane literally surfs along the ground. The condition begins when the tire does not make full contact with the surface and starts to ride up on a film of water. Steering and braking effectiveness decrease. As speed increases, the tire no longer makes contact with the surface.

The speed at which hydroplaning begins is directly related to tire pressure. As tire pressure increases, hydroplaning speed increases. In light airplanes, these speeds vary in the 40- to 55-knot range. Typically, most airplanes touch down well above the minimum hydroplaning speed.

When water is mixed with other substances (oil, rubber, dust, etc.), hydroplaning occurs at a much lower speed. This can happen on the ramp and runup areas, as well as on the runway.

The solution to the hydroplaning hazard is to avoid areas of standing water. Proper runway drainage and grooving help eliminate this hazard. If it is not possible to avoid, land at the slowest possible speed consistent with safety. Proper tire inflation is important.

If snow, ice, or slush are on the runway, aircraft control may be difficult, especially in high winds, with reduced braking efficiency, resulting in a longer than normal ground roll. Without windscreen deice—that's deice, not defrost—a pilot could be faced (pardon the pun) with zero forward visibility during landing. Unless you are Charles Lindbergh flying the *Spirit of St. Louis,* which had a periscope, your only option would be to attempt to scrape off some ice by reaching through the side window.

If there is ice on the windscreen, most likely there is ice on the wings and tail. Arriving pilots may have to contend with adverse flight characteristics, including tailplane icing. Pilots should consider the use of minimum or no flaps under these condition. However, this will result in a high approach speed and longer landing roll.

Regulations require pilots to consider "...runway lengths at airports of intended use...." IFR pilots must consider aircraft landing category. A higher approach speed may place the aircraft in the next category, resulting in higher landing minimums. Snow and snow drifts often render the ILS glide slope inoperative, further raising landing minimums. Pilots operating in this environment must consider these factors when selecting destination and alternate airports or even the advisability of making a flight.

With a clean airframe, a soft-field technique may be best for a snow-, ice-, or slush-covered runway. Use minimum braking, and do not allow the wheels to become locked. Freshly fallen snow causes the wheels to sink to the runway or layer of frozen snow and ice. Notwithstanding the limitations of propeller and wheel pant clearance, a limiting factor would be the radius of the nose-wheel tire, although 50 percent of wheel radius may be a more reasonable factor. Dry snow is easier to deal with than wet snow. In addition,

CASE STUDY
The pilot of a Cessna 206 was briefed about possible icing conditions below 10,000 ft in Bangor, Maine. However, the pilot did not think the icing was significant, even though he could see some ice on the leading edges and the windscreen. The pilot was sure that the ice would be lost at a lower altitude. The ice had different thoughts. The pilot misjudged height because of poor forward visibility. Instead of flying the plane to the ground, as is recommended, the pilot applied the first notch of flaps (10°). The plane stalled immediately, landed on the nose gear, and after one bounce, came to a stop.

The NTSB determined the probable cause as the pilot's improper inflight planning/decision and failure to plan the approach for landing properly so that the airplane would not stall or touch down hard. The accumulation of airframe ice and the pilot's reduced forward visibility were related factors.

with snow, there may be no way to see drifts. Clear glare ice is worst; white, hard-packed snow is not as bad. One of the most difficult situations is patchy ice on the runway where the airplane transitions from one surface to another. When landing on icy runways, there may be no way to stop.

At uncontrolled airports, take the time to circle the field before landing. Look for drifts or other obstacles. Be aware that tracks in the snow on a runway do not ensure a safe landing. Snowmobiles or other vehicles may use the runways, giving pilots the illusion that aircraft have been operating and that the snow is not deep.

Whiteout is an atmospheric optical phenomenon in which pilots appear to be engulfed in a uniformly white glow. Neither shadows, horizon, nor clouds are discernible; sense of depth and orientation are lost. Whiteout occurs over an unbroken snow cover and beneath a uniformly overcast sky when light from the sky is about equal to that from the snow surface. Blowing snow may be an additional cause.

To the VFR pilot, whiteout is disastrous. Snow-covered terrain, an overcast sky, and already reduced visibilities are a strong no-go indicator. At the very first sign, the pilot's only option is a 180° turn to what, hopefully, are better conditions.

The IFR pilot is not immune to whiteout. For example, in the following case, the pilot's first destination did not have an instrument approach. It was snowing, and the pilot reported "whiteout" conditions. The pilot diverted to an airport with an instrument approach.

CASE STUDY

Visibility was reported one-half variable between $1/4$ and 1 mi in snow, indefinite ceiling at 600 ft. Radio contact was lost after the second missed approach. Wreckage was located an hour later.

The following is strictly speculation. Because the crash occurred after the declaration to miss the approach, it appears the pilot decided to miss at the missed approach point. The transition between instrument and visual flight under these conditions is extremely difficult, especially with single-pilot operation. The pilot may have acquired ground contact straight down, but slant range and apparent whiteout conditions would preclude visual contact with the approach environment and airport.

Many commercial operations require two pilots just for this scenario. One pilot stays on the instruments, and the other looks for the airport. I had a similar incident while training an instrument pilot. We were flying at night in rain and fog; ceiling and visibility were at minimums. At the missed approach point, my student started the missed approach. At that point I caught a glimpse of the approach lights. We were certainly not in a position to land and continued the missed approach procedure.

Like most flying activities, extra caution is required at night. Typically, NOTAMs will advertise these conditions, although, it may be necessary to taxi down the entire length of the runway to determine its suitability for takeoff. Make sure there are no snow drifts, and watch out for patchy ice and crosswinds.

It is not unheard of for a pilot, seeing only one set of runway lights, to land on the wrong side of the runway. The pilot has a 50-50 chance,

A low-time pilot said that he could see runway lights prior to the nighttime takeoff on a plowed runway. The pilot aligned the airplane for takeoff between the snow banks along the sides of the runway and by sighting down the runway. The takeoff roll was normal until rotation. As the airplane lifted off, it struck the snow bank on the right side of the runway and nosed over. Airport personnel had plowed the snow at an angle away from the centerline of the runway. The probable cause was the pilot's failure to maintain proper alignment during the takeoff roll. Also sighted as related factors were the dark night, snow-covered runway, inadequate snow removal, and snow bank.

but, a corollary to Murphy's law says that the other guy always has a better than 50-50 chance while you always have worse. If there is any doubt about the landing runway, select an alternate.

The risk assessment and management decision tree can be applied to landings. The following is a list of factors to be considered:

- Airframe ice

- Windscreen ice

- Higher-than-normal approach speed

- Runway surface conditions

- Runway length

Certainly an iced up aircraft and a higher-than-normal approach speed to an ice-covered, short runway results in a no-go decision. What if the runway is clear and long? Here you must rely on your training and experience landing an iced-up airplane. No training or experience: No-go. How about a clean airplane to an ice-covered, relatively short runway? No training or experience: No-go.

En route

During the winter, both VFR and IFR pilots may have to contend with a snow- or ice-covered runway. Airports may be closed due to snow or ice accumulation or removal. Pilots must be sure to check Notices to Airmen (NOTAMs) for airport closures, runway conditions, and breaking action reports. It's also a good idea to recheck en route with a flight service station (FSS) for updates on destination and alternate runway conditions and the status of instrument landing aids. This is especially true if NOTAMs indicate that the airport is in the process of clearing the runways.

> **FACT**
>
> An emergency is simply a situation involving distress and/or uncertainty.

Recall the previous discussion of flight and instrument icing problems. The solution to an iced-over pitot or engine instrument probe is cross-check. Cross-checking all flight and engine instruments should reveal the discrepancy. With the problem diagnosed, a plan can be developed to work around the situation. Loss of flight or engine instruments in most cases renders the aircraft unairworthy. Air traffic control (ATC) must be notified immediately and, most likely, a possible emergency declared. Emergency declaration is a sticky point with some pilots.

All too often pilots find themselves hung up on semantics. What constitutes an emergency? Many pilots think an emergency is an engine failure or inflight fire. An emergency is simply a situation involving distress, the need to resolve uncertainty, or a means of alerting those who are in a position to help. Distress or uncertainty can result from mechanical failure, pilot incapacitation, insufficient fuel, penetration of IFR conditions by a VFR pilot, or a pilot unable to locate the aircraft's position. An emergency exists when declared by the pilot, FAA facility personnel, or officials responsible for the operation of aircraft.

FAA air traffic controllers are trained in various techniques to assist pilots with navigational, mechanical, and weather difficulties. However, none of this training or resources are of use to a pilot unwilling to take advantage of the service.

Thorough flight planning, understanding weather reports and forecasts, and frequent weather updates should alleviate most problems. Mechanical and electronic systems fail, and being human, pilots can find themselves in need of assistance. Confess and communicate the difficulty; certain pilots compound problems by waiting until they have only a few minutes of fuel. A pilot who requests assistance before the situation becomes critical has many more options. Anytime the outcome of a flight becomes uncertain, whether due to unknown position, marginal fuel, weather, or navigational or mechanical problem, contact an FAA facility for assistance. Resolve the situation before a simple flight assist becomes an accident.

A major deterrent to requesting assistance is pilot ego. (Recall the discussion of hazardous attitudes in Chapter 5. Macho certainly fits this category, but invulnerability and even resignation may play a part in a pilot's unwillingness to ask for assistance.) The perceived embarrassment and repercussions of a request for assistance have in many instances turned a bad situation into an impossible one. Fourteen CFR 91.3 specifies the responsibility and authority of the pilot-in-command. Under this rule, the pilot is responsible, and the final authority, for operation of the aircraft. The pilot in an emergency may deviate from any rule to meet that emergency. However, authority does not come without responsibility. The pilot may be required to send a written report of any deviation.

> **CAUTION**
> Do not blindly follow instructions. This may compound the problem. Keep in mind that you are still pilot-in-command. If you are flying VFR and penetrating clouds is unavoidable, inform ATC immediately.

Pilots who request assistance usually do not deviate from CFRs. However, their actions may be reviewed by the FAA. With few exceptions, this results in counseling, unless there is evidence of excessive pilot deficiencies. Flight assistance service cannot be used continually by a few pilots who refuse to obtain or maintain basic proficiency. As stated in the *Aeronautical Information Manual,* "When you are in doubt of your position, or feel apprehensive for your safety, do not hesitate to request assistance. The FAA's Air Traffic Service facilities are ready and willing to help." A final thought: You always want to be in a position to represent yourself at the NTSB hearing.

The following factors should be considered during any emergency situation:

- Federal Aviation Regulations.

- Divert to an alternate.

- Return to the departure airport.

- Weather.

- Training and experience.

- Missed approach.

Regulations specify required pilot actions during loss of flight, engine, or navigation instruments. The first consideration may be to divert to a suitable alternate. After all, you know the nearest area of clear weather, don't you? If you have just departed, the safest solution may be to return to the departure airport. You may need updated weather to determine the most practicable alternative. Whatever the solution, it must be based on your training and experience. For example, if you are not comfortable with a partial-panel approach, possibly to minimums, you need to develop an alternate plan.

With an iced-over pitot tube due to equipment failure or lack of pitot heat, the result is loss of the airspeed indicator. Solution: Fly attitude and power. You should be familiar with the attitude and power settings required for climb, cruise, descent, and approach. If you are not, learn them.

What about loss of the static ports? This renders all pitot-static instruments inoperative. Solution: Select the alternate static source. (You did remember to check it before flight?) You do not have an alternate static source? What are you doing in icing conditions? As a last resort, consider breaking the glass on the vertical speed indicator. Some pilots actually carry a small hammer just for this purpose.

Without pitot heat and an alternate static source, carefully consider the risk of flying in or near icing conditions. Should an emergency, incident, or accident occur, do not be surprised that an inspector recommends a "careless or reckless operation" violation.

Instrument pilots are taught and tested on flying the aircraft with loss of some gyroscopic instruments. Unfortunately, after certification, many pilots fail to maintain this skill. Have you ever lost the vacuum pump in IFR conditions or on a dark, moonless night? I have. I have made two no-gyro approaches, one to ILS minimums. The point is that it can and does happen. Solution: Maintain partial-panel flying skills. This includes a no-gyro approach to minimums and a missed approach. For myself, during these occurrences, I employed the strategy of executing a missed approach if the localizer needle neared full deflection or at any point the airplane descended below the glide slope.

Climbout

On climbout, keep a close watch on cylinder heat temperature. Because of restrictions—baffles—to cooling airflow installed for cold weather operations and the possibility of temperature inversions, it is possible to overheat the engine at normal climb speeds. If the cylinder head temperature nears the critical stage, increase airspeed or open cowl flaps or both.

Most aircraft certified for flight in known icing have a published minimum speed for flight in icing conditions. This speed is designed to prevent an angle of attack that allows ice to build up on the underside of the wing or aft of the ice-protected area. This airspeed is typically not the best rate of climb speed; often it is much higher. In fact, in some cases this speed may be so high that it essentially prohibits climbs in icing conditions.

Recall from Chapter 2 that the Commander 114 AFM states: "Accumulation of ice on unprotected lower surfaces is minimized by maintaining a minimum airspeed of 110 KIAS...." What if you cannot climb at this airspeed? Then icing conditions are too severe for flight. It is time to go to an alternate option. This may be a lower altitude, a 180° turn, or

FACT Airspeed decreases and stall speed increases as ice accumulates.

flight to an alternate airport. Aircraft not certified for flight in known icing have no such recommended icing climb speeds. As mentioned previously, the pilot of a noncertified aircraft becomes a test pilot in icing conditions. Should a climb be necessary, the pilot has two choices: Climb to get through the icing as quickly as possible or climb at as high an airspeed as possible to minimize exposing the lower surface of the wing to increased icing. As ice accumulates, airspeed decreases and stall speed increases. Pilots must very carefully consider a climb through icing conditions. Turbocharged airplanes have a significant advantage over airplanes with normally aspirated engines. The following factors must be considered:

- How thick is the icing layer?

- What are the tops?

- Can you easily get on top?

- Do you have an out?

If the answer to any of these questions is no, decision: No-go.

Updating Weather En route

A pilot's responsibility does not end with an understanding of weather and forecasts and a complete preflight briefing. Because of the dynamic character of the atmosphere, data must be updated continually. Surprisingly, many pilots have not been taught or have not learned the importance of updating weather reports and forecasts. Failure to exercise this pilot-in-command prerogative, as we have seen, can have disastrous results.

Updating weather en route begins with the "complete picture." In fact, many pilots follow weather trends for several days before a planned flight. In this way they get a feeling for the weather. Without this background knowledge, they will not be able to fully apply the weather briefing and updated weather en route. For example, en route a pilot obtains information that the destination is below his or her minimums. The next question is: Why? Is it due to the delayed improvement of stratus clouds and fog, a faster than forecast advance of a frontal system, or the approach of a hurricane. The pilot needs to know in order to develop a sound alternate plan.

For effective risk management and decision making, updates must be obtained far enough in advance to be acted on. This must be done before critical weather is encountered or fuel runs low. Arriving over a destination that has not improved as forecast or has deteriorated is folly. At the first sign of unforecast conditions, Flight Watch should be consulted, and if necessary, an alternate plan developed. This may mean an additional routine landing en route, which is eminently preferable to at best a terrifying flight or at worst an aircraft accident.

Up-to-date and accurate information and a knowledge of how to apply it are key to intelligent inflight risk assessment and management and decision making. There are five sources of information:

- The pilot
- Flight service stations (FSSs)
- En route flight advisory service (Flight Watch)
- Air traffic control facilities—centers and towers
- Automated observation broadcasts

The pilot, through direct observation and analysis of aircraft instruments, is a key source. It was already mentioned how the pilot is often the only means of evaluating inflight weather conditions, especially icing and turbulence, cloud types, and winds and temperatures aloft. To apply the information successfully requires a sound knowledge of aviation weather phenomena and theory.

A pilot is flying along VFR or IFR and encounters freezing precipitation. In freezing precipitation, aircraft without a heated pitot and alternate static source, especially in IFR conditions, would be in serious trouble. Another significant factor, especially for aircraft without ice protection equipment, is that accumulated ice could be carried all the way to the ground, making landing extremely hazardous. It cannot be overemphasized that this hazard can affect VFR as well as IFR operations.

Should freezing precipitation be encountered in aircraft without ice protection equipment, virtually the only option, and certainly the safest, is to divert to an area of warmer air. Since it's not logical that a pilot would take off in freezing precipitation, the aircraft must have come from a freezing precipitation–free area. VFR, the pilot may be able to descend to warmer air below; otherwise, the pilot should reverse course immediately, fly to warmer air, and reevaluate his or her options. IFR, the pilot may have one additional option—climb to warmer air above. This option must be evaluated within the context of the icing options discussed thus far. Do you know the level of the warmer air above; can you get on top; do you have the performance to climb in icing conditions? Unless you have affirmative answers to these questions, the decision is no-go.

FAA flight service stations (FSSs) are a primary source of weather information. All the information available during the preflight weather briefing is accessible inflight through FSS communications outlets. This information includes current observations, PIREPs, real-time radar reports and satellite imagery, and often, update forecasts.

Chapter 5 touched on the problem of knowing when to call it quits and land or turn around. Unfortunately, a pilot's training may do him or her a disservice. Typically, the instructor puts the student under a hood and simulates just entering a cloud. This certainly is not the behavior an instructor wants to instill in pilots! The goal is to have the pilot turn around before entering less than VFR conditions.

I encountered just such a situation on a flight from the San Francisco Bay area to Salt Lake City. It was October, but an upper level low was over the Great Basin. The flight was uneventful to Tonopah, Nevada, where I refueled. The weather into Salt Lake City was marginal. I dodged rain showers and had to fly through the passes. Approaching Delta, Utah, I updated the weather into Salt Lake City with Cedar City Radio. MEAs were high and the freezing level low. An IFR flight was no-go. The mountains were obscured by clouds and showers, with no improvement expected. With rising terrain between Delta and Salt Lake City, the risk was too high, and I landed short at Delta. I rented a car for the rest of the trip. Could I have flown into Salt Lake City? Probably not. The risk was too high. My only safe option was to land at Delta.

Selected FSSs provide a continuous broadcast of weather advisories and urgent PIREPs over selected radio navigation aids (VOR). This service is known as the Hazardous Inflight Weather Advisory Service (HIWAS). When a weather advisory affects an area within 150 mi of a HIWAS outlet, an alert is broadcast once on all frequencies, except Flight Watch and emergency. Pilots are instructed to monitor HIWAS or contact flight service or Flight Watch.

The purpose of Flight Watch is to provide pilots with timely and meaningful weather advisories. Flight Watch can provide information on cloud tops, temperatures aloft, reported and forecast icing, and current surface conditions. En route Flight Advisory Service (EFAS), radio call "Flight Watch," is a special function provided by selected FSSs. Flight Watch provides meteorologic information for that phase of flight that begins after climbout and ends with descent to land. Flight Watch is a central point for the collection and dissemination of pilot weather reports. The effectiveness of Flight Watch, to a large degree, depends on this two-way exchange of information.

Are updated reports consistent with the forecast? If not, why? Flight Watch Specialists, through their training, are in an excellent position to detect forecast variance. Whether the forecast was incorrect or conditions are changing faster or slower than forecast, the pilot needs to know and plan accordingly. Flight Watch is in the best position to provide the latest information and suggest possible alternatives. Without this information, a pilot is in a poor position to evaluate and manage risk or make sound decisions. .

Flight Watch is not intended for flight plan services, position reports, initial or outlook briefings, aeronautical information, or single or random weather reports or forecasts. This information should be obtained through normal FSS communication channels.

Flight Watch communications, below 18,000 ft, are on the common frequency of 122.0. Since Flight Watch has a number of transmitter sites on the common frequency, it is very important for pilots to state the aircraft's position on initial contact. For example, "Oakland Flight Watch, Cessna seven three five two juliett, Lake Tahoe, over." If you're not sure of which Flight Watch you're calling, just call Flight Watch and state the aircraft's position, "Flight Watch, Champ four three three zero charlie, Big Sur, over." (OK, that was the commercial. I would be severely criticized by my colleagues at Flight Watch if I didn't mention it. Pilots' failure to provide position on initial contact is the biggest complaint from Flight Watch specialists.)

At and above flight level 180, each Air Route Traffic Control Center (ARTCC) has its own discrete Flight Watch frequency. These frequencies are shown in Fig. 6-3.

Even single-pilot IFR operations can use Flight Watch services. I have never had a problem with en route controllers not allowing me off frequency for a few minutes to update weather. If you are close to a sector boundary, the controller may have you wait until you are in contact with the next sector. Do not try this in congested terminal airspace. Again, the key is to update information far enough in advance to allow effective risk management and decision making.

A weak front was moving through central California following Thanksgiving. The system brought occasional IFR conditions, mountain obscuration, and icing above about 10,000 ft. The MEA across the coastal mountains was 7000 ft. Applying the risk assessment and management decision tree, my IFR flight was a go in the Cessna 172.

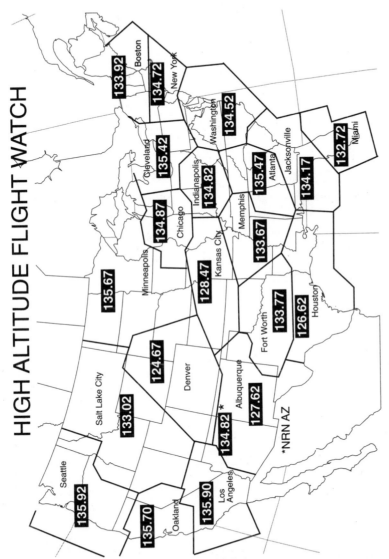

HIGH ALTITUDE FLIGHT WATCH

Fig. 6-3. Each Air Route Traffic Control Center has a discrete Flight Watch frequency for use at and above flight level 180.

I initially filed for 6000 ft, well below the freezing level. Between Priest and Panoche, over the coastal mountains, the MEA climbs to 7000 ft. Center cleared me to 8000 ft. With pitot heat on, I climbed to the new altitude. I carefully monitored outside air temperature and for any indications of structural or induction ice. At 8000 ft the temperature read $+2°C$, and the precipitation remained liquid. At this point I observed the scene in Fig. 6-4.

I coordinated with center to "leave the frequency for no more than 3 minutes." My request was granted, and I contacted Flight Watch. Destination weather remained a go, and I provided a PIREP. Some pilots recommend providing PIREPs directly to center or tower controllers. Certainly they would be interested in significant icing within their airspace. Unfortunately, many useful PIREPs never make it beyond individual sector controllers. It is doubtful if my report would have been forwarded. Why? I had nothing to report that would affect the controller's operation, whose primary function is separating traffic. However, my report would be extremely valuable to pilots

Fig. 6-4. It's important to check weather before encountering hazardous flight conditions.

planning flights in this area. I hope you can all see how valuable a PIREP can be in this situation—in clouds, no turbulence, no ice, temperature above freezing. Therefore, I strongly recommend passing PIREPs directly to Flight Watch.

ATC specialists at towers and centers are another resource for inflight weather information. Many towers provide Automatic Terminal Information Service (ATIS) with local weather information and weather advisories. Radar facilities can assist with weather avoidance, but remember, this is not their primary job, and ATC radar is not designed specifically to detect weather. The FAA, however, is in the process of updating ATC displays to better depict convective weather. This will provide the controller with Doppler weather radar information. It will not be a substitute for onboard weather avoidance equipment, however.

With the increased availability of automated weather observation systems, AWOS/ASOS has become a significant weather resource. This is especially true for uncontrolled fields. By monitoring AWOS/ASOS, as well as sky conditions, visibility, and altimeter setting, pilots receive surface wind information. From this they often can determine favored runway. By checking the *Airport/Facility Directory*, traffic direction can be determined for that runway. This allows pilots to more effectively enter the traffic pattern.

The following case study does not relate directly to icing, but it is a good example of using the system—FSSs and ASOS—for evaluating and managing risk and the decision-making process. You may ask why I used FSS and not Flight Watch? Since flight plan revisions were required, Flight Watch was not appropriate. Also, I only needed specific weather reports. I was able to accomplish both tasks with one radio contact.

Descent and Approach

Prior to beginning descent, where possible, obtain destination weather and NOTAMs. This can be accomplished through ATIS, FSSs,

> **CASE STUDY**
>
> The Weather Channel reported a fast-moving cold front, with strong surface wind, approaching Palm Springs. Because of the large amount of traffic for the AOPA convention, no tie-downs were available. With my business concluded, I decided to leave on Saturday. I received a briefing and filed a VFR flight plan to Lancaster's Fox Field—plan A.
>
> It was a beautiful flight, smooth air with a high overcast. With plenty of fuel, I decided to continue on to Harris Ranch in the San Joaquin Valley. I updated weather and revised my flight plan with Riverside Radio—plan B.
>
> Over Bakersfield, I made a position report to Rancho Radio and received updated weather. Navy Leemore, adjacent to Harris Ranch, was reporting visibility three-quarters of a mile in blowing dust, wind gusting to 35 knots! Well, I knew where the front was. I checked the weather at Visalia and again amended my flight plan—plan C.
>
> As I proceeded toward Visalia, I could see the billowing dust storm approaching from the northwest. Fortunately, Visalia had an ASOS. I was only 10 mi south of Visalia when the ASOS reported the arrival of the storm. I contacted Rancho Radio, advised them of the situation, canceled the flight plan, and descended into the traffic pattern at Tulare—plan D. I got the airplane tied down about 5 minutes before the storm hit.

and AWOS/ASOS. (Automated weather observations do not carry NOTAMs.) With this information you can determine landing runway, approaches in use, and surface conditions. Should the field be unsuitable for landing, this is the time to begin selecting an alternate. Remembering that the goal is to minimize exposure to icing; you do not want to descend and execute an approach, only to find that you have to climb back to altitude and proceed to an alternate. In order to minimize exposure, negotiate with ATC to obtain a continuous descent. Avoid, if possible, level flight in clouds. ATC is usually very responsive to such requests.

During descent and approach, you want to avoid high angles of attack. Like climbout, high angles of attack in icing conditions may allow ice to form aft of the ice-protected area under the wing. Use the autopilot with caution. In icing conditions the autopilot altitude hold is going to do just that, hold altitude, often at the expense of increased angles of attack. This is especially true if holding should be required.

On airplanes equipped with a float-type carburetor, the use of carb heat is normally recommended during reduced (out of the airspeed indicator's green arc) or closed-throttle operation. Since the heat is generated by the running engine, during extended periods of low-throttle operation, it may be desirable to advance the throttle periodically to ensure proper engine operation. On airplanes with a pressure carburetor—much less susceptible to icing—carb heat may only be recommended if carburetor icing conditions exist.

Remove carburetor heat prior to a go-around or balked landing. With carburetor heat, loss of power may be critical at low altitude and low airspeed, and engine damage may occur with full heat on a go-around at takeoff power on some powerplants.

> **CASE STUDY**
> The following is an all too common scenario in NTSB accident reports. A student pilot, flying a Cessna 172, took off to practice touch-and-go landings at a nearby airport. On returning 1 hour later to Will Rogers World Airport, the engine quit. The airplane was about 5 miles west of the destination. The pilot restarted the engine, but it quite again about 1 mile west of runway 35L. The airplane crashed in a field short of the airport. According to the carburetor icing probability chart, weather conditions at the time of the accident were conducive to the formation of carburetor ice. Additionally, the engine was test run and found to have no mechanical anomalies. After the accident, the pilot told investigators that carburetor heat was never used.

It is easy to get lulled into a false sense of security. I have done most of my flying in the Southwest, where carburetor ice typically is less of a problem due to the dryer climate. Flying various airplanes, most less susceptible to carb ice than the Continental O-200 in the Cessna 150,

I had become very lax in the use of carburetor heat. During 1997–1998, I spent a year in the Washington, D.C. area. Talk about culture shock, or maybe it should be called climate shock. With the help of some excellent flight instructors at Montgomery Aviation, in Gaithersburg, Maryland, I became reacquainted with the proper use of carburetor heat. I shall not forget this lesson.

Postflight Considerations

The flight is not over until the aircraft is parked, tied down, and secured. In addition to ensuring that all switches are off, securing the controls, and chalking and tying down the aircraft, there are some additional precautions for cold weather operation.

The following should be considered before leaving the aircraft. You might want to add them to your postflight checklist.

1. During engine shutdown, consider turning off the fuel to run the carburetor dry.

2. As soon as possible, fill the tanks with fuel, especially if the aircraft is going to be left in a heated hanger.

3. If the aircraft is to be left outside, install engine and pitot covers.

4. If the aircraft is going to be flown again, place a blanket over the engine.

5. If the forecast is for snow or clear and colder temperatures, install rotor or wing covers.

6. If the aircraft is equipped with an oil dilution system, consider diluting engine oil.

Shutting off the fuel during shutdown lessens the hazard of fire during preheat. As recommended for normal operation, fuel the aircraft when it is left overnight. There is a greater chance of water accumulation in the fuel tanks when the aircraft is left in a warm hanger. Since pitot, engine, and inlet covers can be hidden by snow, add prominent red flags to signify their presence. These must be removed before flight. The removal of each cover should be on the preflight checklist. If the airplane is going to be flown again the same day, you may want to consider placing a blanket over the engine to conserve as much heat as possible. Be careful! The engine, and especially the exhaust stacks, will be very hot. Wing and rotor covers will prevent snow and frost from adhering to these surfaces. This will greatly facilitate preparation of the aircraft for a morning departure. An oil dilution system greatly enhances starting in cold weather. However, be sure to following the manufacturer's recommendations.

Beginning with a warm aircraft is always preferable to preheat and deicing. It is a good idea to check on the availability of a hanger, along with preheat and deicing facilities, prior to beginning a flight that requires remaining overnight. You certainly would not want to wind up at an intermediate stop only to find that there are no such services. The best procedure would be to check with airport operators prior to the trip when cold weather is expected.

Putting It Together

Let's apply our knowledge of icing to some actual flight situations. The following weather event occurred on December 21, 1998. We will begin with a flight from Goodland to Wichita, Kansas.

Recall the risk assessment and management decision tree (Fig. 5-6) from Chapter 5. The first step is planning. A review of aeronautical charts and the *Airport/Facility Directory* or *Terminal Procedures Publication* reveals the following:

Goodland: Elevation 3656 ft
                         RWY 12-30 5419 ft
                         RWY 17-35 1800 ft (turf)
                         RWY 5-32 3501 ft
Wichita: Elevation 1332 ft; magnetic variation 7°E.
                         RWY 1L-19R 10,300 ft
                         RWY 1R-19L 7302 ft
                         RWY 14/32 6301 ft

Terrain:	Relatively flat, sparsely populated prairie—except for the cities—sloping from about 3500 ft in western Kansas to about 1500 ft around Wichita
Altitudes:	VFR, 500 to 1500 ft AGL and up; IFR, minimum altitudes 5000 to 3000 ft
Environment:	Daylight, midday flight, departing around 18Z

We have the capability of flying VFR or IFR and can select from a number of airplanes. We can take a Piper Warrior or an ice-protected Commander 114. The choice will depend on the weather.

A pictorial view of the weather is often helpful in developing the "complete picture." For this we need access to graphic products. With FAA and NWS consolidation, a visit to an aviation weather facility is often not practical. However, with DUATs and the Internet, these products are becoming more accessible. Obtaining graphic products prior to the standard briefing provides a general picture of the weather. Like checking terrain, altitudes, and airport information, a preliminary look at the weather provides a helpful background for the preflight briefing.

Figure 6-5 contains a morning weather depiction and radar summary chart for December 21, 1998. An arctic cold front extends southeast of the route. The general weather along the route: departure VFR, en route MVFR, and destination IFR.

Freezing drizzle in southeast Kansas is a red flag for supercooled large droplets (SLDs). Station models show freezing drizzle behind the front from the Texas panhandle to Missouri. Recall the discussion of the probable locations of SLDs in Chapter 1. This location coincides with the location of SLDs 25 to 130 mi behind the arctic front, exactly where we would expect this phenomenon. The arctic cold front is undercutting and lifting warmer air to the south. This feature is often referred to as *overrunning warm air aloft*. Precipitation falls as liquid into colder air below and then freezes on contact with a surface that is below freezing.

With our knowledge of weather patterns, we would expect improvement from the north during the day—we will verify this with the

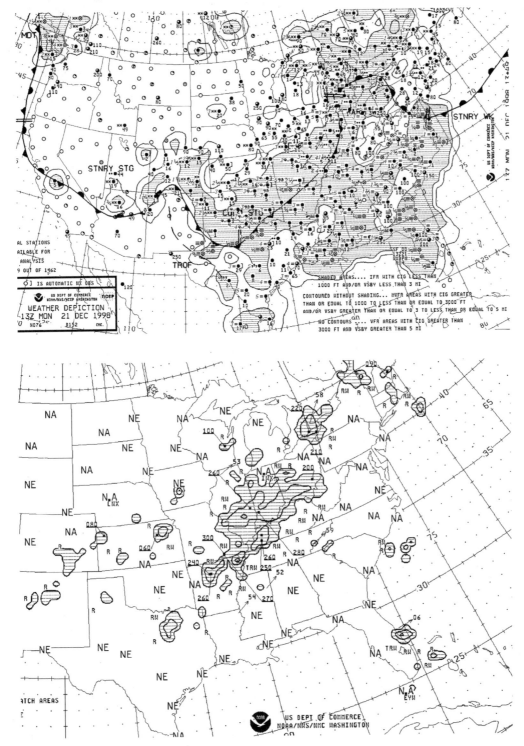

Fig. 6-5. A pictorial view of the weather is often helpful in developing the "complete picture."

Area Forecast and TAFs. If we are looking for alternates, north appears to be the best bet. Upslope continues over the plains of Colorado, with snow and IFR conditions—not favorable for an alternate. Widespread areas of freezing drizzle and IFR conditions continue south and east of our route. Even though we may find legal alternates in these areas, they may not be the best choice. Legal does not necessarily mean safe.

The radar chart is encouraging. It depicts scattered light to moderate precipitation, in the form of rain or rain showers. Rain indicates a stable air mass—typically less serious icing. Tops of precipitation are in 6000- to 8000-ft range. With relatively stable air, we would expect cloud tops to be within several thousand feet of radar tops, in the 9000- to 12,000 ft range. Since icing tends to be more serious in cloud tops, based on this information, we would want to avoid the 9000- to 12,000-ft altitude range. No convective activity is reported for the route. What about precipitation in the vicinity of Wichita? The radar chart shows the symbol *NA*. For some reason, the data are not available. With no echoes (*NE*) reported in Oklahoma and the weather depiction showing light, steady precipitation, we could reasonably expect the conditions shown in western Kansas to extend into the Wichita area. Note that the front is most active from northern Arkansas through the Ohio Valley.

Satellite imagery in Fig. 6-6 confirms our interpretation of the weather depiction and radar summary charts. There are widespread clouds but relatively low tops with no convective activity along the route. What about the clouds over Nebraska? The weather depiction shows mostly clear. These are most likely automated observations, even though not depicted as such, which report maximum cloud heights of 12,000 ft. The visible image shows relatively thin clouds; the IR image shows cold tops. This is most likely a cirroform layer.

With access to the Internet, we can check the experimental icing products. The neural network icing product, surface to 6000-ft composite, is shown in Fig. 6-7. Moderate to severe icing is projected for northern Texas, most of Oklahoma, and through southeastern Kansas, Missouri, and Illinois. These conditions are expected along and behind the cold front, where we would expect them. Light to

IR SATELLITE IMAGE

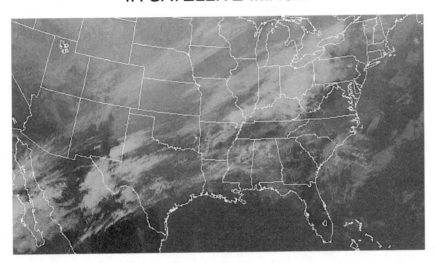

VISIBLE SATELLITE IMAGE

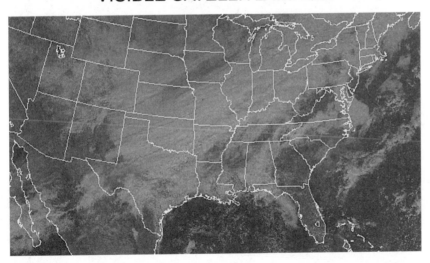

Fig. 6-6. These satellite images indicate widespread clouds but relatively low tops, with no convective activity along the route from Goodland to Wichita.

moderate icing is forecast for western Kansas. Again, what we would expect in the colder air to the north.

Figure 6-8 contains the 6000- and 9000-ft stovepipe icing potential projections. At the 6000-ft level, moderate potential exists over western Kansas and a high potential over southeastern Kansas, in the

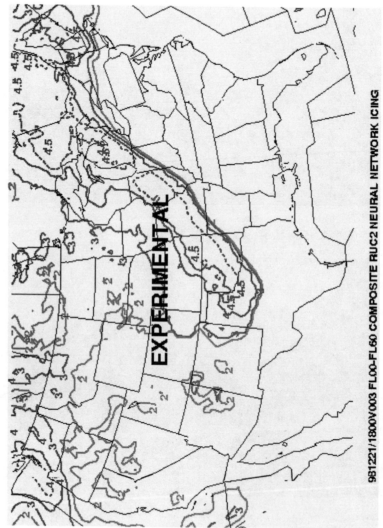

981221/1800V003 FL00-FL60 COMPOSITE RUC2 NEURAL NETWORK ICING

Fig. 6-7. The Neural Network Icing model predicts light to moderate icing for western Kansas, becoming more severe around Wichita.

9,000-FOOT ICING POTENTIAL

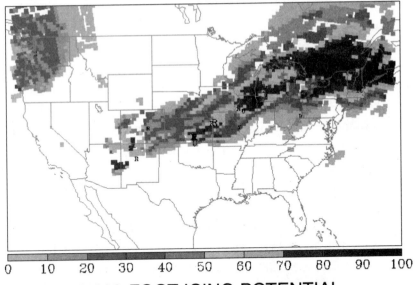

0   10   20   30   40   50   60   70   80   90   100

6,000-FOOT ICING POTENTIAL

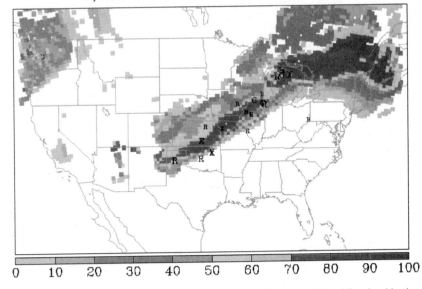

0   10   20   30   40   50   60   70   80   90   100

Fig. 6-8. Not surprisingly, the Stovepipe model predictions for icing potential agree with the Neural Network model, and are consistent with the weather advisories.

same areas depicted by the neural network model. The 9000-ft level predicts moderate potential over Kansas and isolated areas of high potential in the Southeast, with the most serious icing at lower levels. Again, this is what we would expect.

From Fig. 6-9, icing tops for Kansas are forecast in the 15,000- to 20,000-ft range. That is pretty high, but remember, the model cannot take all factors into consideration. Also, from Fig. 6-9, SLD tops are expected to be in the 5000- to 10,000-ft range. (I know you cannot see the color; you will just have to take my word.) We can conclude from these models that the best escape route, again, would be to the north.

With the preceding background, we are ready for the briefing from either an FSS or DUATs. Each will contain essentially the same information. We will call the FSS or log onto the computer, provide the necessary background information, and request a standard briefing. The DUATs briefing will appear in the following sequence.

The first product displayed is the Area Forecast:

```
CHIC FA 211045
SYNOPSIS AND VFR CLDS/WX
SYNOPSIS VALID UNTIL 220500
CLDS/WX VALID UNTIL 212300...OTLK VALID 212300-220500
ND SD NE KS MN IA MO WI LM LS MI LH IL IN KY

SYNOPSIS...10Z LOW PRES NRN LM WITH CDFNT TVC-SBN-ARG.
HI PRES NERN SD. BY 05Z CDFNT OVR XTRM ERN KY. HI PRES SERN
KS. LK EFFECT SHSN/BLSN OVR GRTLKS THRU 05Z.

KS
SERN...CIG OVC010-020 TOP 120. OCNL VIS 3-5SM -FZDZ BR. 15-
17Z FZDZ BECMG SN. 21Z AGL SCT-BKN020 CIG OVC040 TOP 080.
OTLK...VFR.
WRN...CIG BKN010-020 TOP 120. VIS 3-5SM -SN. 18Z AGL SCT-
BKN020 CIG BKN040. OCNL -SN. 22Z AGL SCT-BKN040 BKN100 TOP
150. OTLK...VFR.
CNTRL/NERN...CIG BKN-SCT010-015 OVC030 TOP 100. OCNL VIS
3-5SM -SN. 20Z AGL SCT025 CIG BKN040 TOP 100. OTLK...VFR.
```

ICING LAYER TOPS

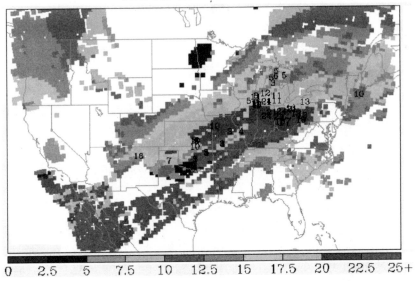

0   2.5   5   7.5   10   12.5   15   17.5   20   22.5   25+

SLD LAYER TOPS

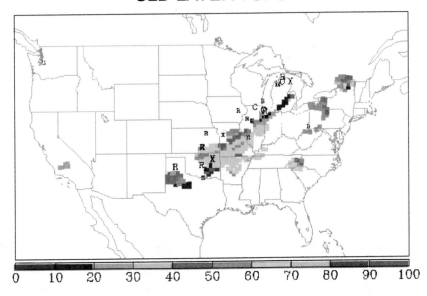

0   10   20   30   40   50   60   70   80   90   100

Fig. 6-9. Remember that icing models cannot take all factors into consideration; they are not a substitute for current weather advisories.

The synopsis discloses a cold front east of our route in southeastern Missouri, moving eastward, and high pressure over northeastern South Dakota. The synopsis confirms our analysis of the weather depiction and radar summary charts. The front is not mentioned in Oklahoma and Texas because it is relatively weak and defuse, and areas are not within the Chicago FA coverage. The front is, however, strongest in the Ohio Valley and Arkansas, even producing some thunderstorms. The front is forecast to move eastward during the period. High pressure is expected to move into southeastern Kansas. This confirms our expectation of better weather to the north.

The route forecast indicates for western Kansas scattered to broken clouds at 2000 ft AGL, ceilings at 4000 ft and broken, tops at 12,000 MSL, and occasional light snow until middle to late afternoon; central Kansas 1000 to 1500 ft AGL scattered to broken, 3000 ft AGL overcast, tops at 10,000 ft, and visibility occasionally 3 to 5 mi in light snow; southeastern Kansas 1000 to 2000 ft AGL, tops at 12,000 ft MSL, and visibilities occasionally 3 to 5 mi in moderate freezing drizzle becoming snow by 17Z, with conditions improving during the afternoon. Note how the forecast very closely agrees with our analysis of the experimental icing product and satellite images. The forecast confirms our analysis that in general we will want to avoid altitudes near the cloud tops, the 9000- to 12,000-ft range, unless we can remain clear of clouds.

Weather advisories are depicted in Fig. 6-10. AIRMET ZULU pertains to the second two-thirds of the flight; occasional moderate mixed or rime icing is predicted in clouds and precipitation below 18,000 ft, freezing level at the surface. This is consistent with the general icing tops predicted in the stovepipe model, icing layer tops (see Fig. 6-9). Why no icing in northwest Kansas? The air is too cold. Cloud droplets and precipitation would be frozen. The AIRMET does, however, imply possible SLDs (*MXD ICGICIP*) within its area of coverage.

The AIRMET refers us to SIGMET PAPA. The SIGMET, represented by the gray-shaded area in Fig. 6-10, warns of

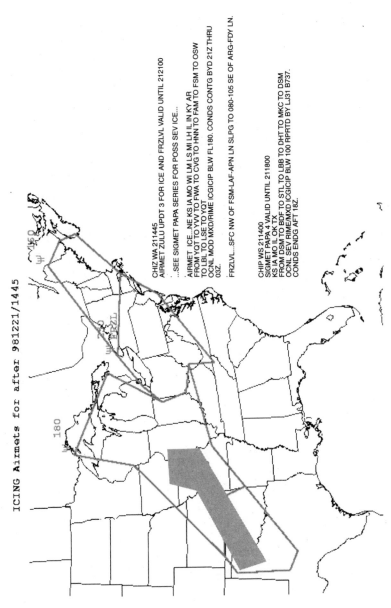

ICING Airmets for after 981221/1445

CHIZ WA 211445
AIRMET ZULU UPDT 3 FOR ICE AND FRZLVL VALID UNTIL 212100

...SEE SIGMET PAPA SERIES FOR POSS SEV ICE...

AIRMET ICE... NE KS IA MO WI LM LS MI LH IL IN KY AR
FROM YQT TO YVV TO FWA TO CVG TO HNN TO FAM TO FSM TO OSW
TO LBL TO LSE TO YQT
OCNL MOD MXD/RIME ICGICIP BLW FL180. CONDS CONTG BYD 21Z THRU
03Z.

FRZLVL...SFC NW OF FSM-LAF-APN LN SLPG TO 080-105 SE OF ARG-FDY LN.

CHIP WS 211400
SIGMET PAPA 4 VALID UNTIL 211800
KS IA MO IL OK TX
FROM DSM TO BDF TO STL TO LBB TO DHT TO MKC TO DSM
OCNL SEV RIME/MXD ICGICIP BLW 100 RPRTD BY LJ31 B737.
CONDS ENDG AFT 18Z.

Fig. 6-10. The gray shade graphically depicts the area described in Chicago SIGMET PAPA 4.

occasional severe rime or mixed icing in cloud and precipitation below 10,000 ft—a definite indicator of SLDs. Again, this corresponds to the stovepipe model SLD layer tops (see Fig. 6-9). However, the SIGMET states that this condition will be ending after noon (*CONDS ENDG AFT 18Z*).

Current weather:

METAR KGLD 211653Z 35005KT 7SM OVC040 M22/M27 A3034

METAR KICT 211655Z 360020KT 2 1/2SM -SN OVC016 M13/M15 A3027

Current weather at Goodland is well above the lower limit of the weather depiction/FA VFR category. Surface temperatures are very cold, and wind is not much of a factor. We will have to make sure that there is no snow or ice on the airplane before takeoff. Some type of preheat should be used. Wichita is below FAR basic VFR, with light snow falling and winds out of the north at 20 knots, which could present a problem landing.

Pilot reports:

DDC UA /OV DDC/TM 1818/FLUNKN/TP SW4/RM BA FAIR-POOR/NEG ICE DURGD

TOP UA /OV KFOE/TM 1732/FL140/TP FA20/TA 0/IC LGT-MOD MXD 025-140/RM FM ZKC

ICT UA /OV ICT/TM 1809/FLUNKN/TP BE9L/SK OVCUNKN-TOP095/IC TRACE/RM IC DURGC

ICT UA /OV KICT 15SE/TM 1659/FLUNKN/TP DC9/SK OVCUNKN-TOP092/TA 00 AT 092/IC MDT RIME 050/RM DURGC

At Dodge City (*DDC*), southeast of Goodland, in western Kansas, a metroliner reports braking action fair to poor and no ice during descent. This implies a cloud layer, but what are the tops? Unfortunately, this is an example of an incomplete pilot report. Dodge

City is just west of the AIRMET area. At Topeka (*TOP*), in northeast Kansas, a Falcon jet reports light to moderate mixed icing from 2500 to 14,000 ft, well within the AIRMET area. This report was made through the Kansas City Center's CWSU (*ZKC*), but again, cloud tops are missing.

The two Wichita pilot reports are extremely helpful. They report cloud tops at 9200 and 9500 ft. The King Air reports a trace of icing during climb; the DC9 experienced light to moderate rime ice at 5000 ft during climb. These reports confirm that the most serious icing has moved to the southeast.

Terminal Aerodrome forecasts:

```
TAF KGLD 211722Z 211818 33010KT 3SM -SN BKN015 OVC140
TEMPO 1820 P6SM SCT015 OVC140
FM2100 34008KT P6SM FEW015 SCT140 SCT250
FM0100 32006KT P6SM FEW250

TAF KICT 211722Z 211818 36015G25KT 3SM -SN BR OVC025
TEMPO 1823 1SM -SN BR OVC015
FM2300 35015G25KT P6SM OVC025 TEMPO 2301 SCT015 BKN090
FM0100 33008KT P6SM SKC
```

The Goodland TAF reveals improving conditions throughout the day. (Note that the departure TAF is not normally part of an FSS briefing—except on request.) Goodland would be a good retreat plan should we run into problems. The Wichita TAF also indicates improvement, but with much lower conditions and slower improvement than expected at Goodland. IFR conditions are expected to continue through late afternoon, with northerly, gusty surface winds. Frozen precipitation is predicted through late afternoon. Decreasing frozen precipitation is a positive indicator of lessening serious icing potential. The light snow will have some effect on surface conditions. This needs to be monitored throughout the flight. The weather is expected to clear during the evening.

Wind and temperatures aloft forecast.

	3000	6000	9000	12,000
GLD		3310	3418-25	3523-31
ICT	3115	3018-15	3921-21	2825-27

Winds generally are northwesterly, averaging 15 to 25 knots and increasing with altitude. Temperatures aloft are very cold. Above 9000 ft the air is too cold for serious icing in stratiform clouds. This confirms our interpretation of no icing in western Kansas—downward vertical motion, cold temperatures, and droplets in the form of ice crystals.

Notices to Airmen:

!GLD 12/005 17-35 CLSD
!GLD 12/006 5-32 CLSD
!GLD 12/007 12-30 1 IN LSR

!ICT 12/020 14-32 CLSD
!ICT 12/026 1L-19R 1/2 IN SIR
!ICT 12/027 1R-19L 1/2 IN SIR

The shorter runways are closed at both Goodland and Wichita. This is typical with a storm system moving through the area. Primary runways are cleared first and then air carrier taxiways and ramps. We will use extra caution taxiing because taxiways and ramps, especially for general aviation, typically are not cleared as rapidly as those used for air carrier operations.

Let's analyze an IFR flight in an ice-protected airplane.

Takeoff: Runway 12-30 at Goodland has 1 in of loose snow on the runway. Since its length is 5000 ft, we will select a marker halfway down the runway; should we not be airborne, out of ground effect, with a positive rate of climb at this point, we will abort the takeoff. Decision: Go.

En route: We have two options: climb to the top or select an IFR altitude below about 9000 ft. We want to stay out of the tops as much as possible. Decision: Go. We will check en route for updates on weather conditions with Flight Watch. Should we wish to update surface conditions at Wichita, we will have to contact an FSS. If conditions should deteriorate, we will return to Goodland.

Landing: We will plan to stay at our cruising altitude until we can make a continuous descent for landing. Runways at Wichita have $1/2$ in of snow and ice on them. Taking into account that TAF surface winds are true and runways are numbered magnetic, from Fig. 5-3, "Crosswind Component," we calculate a crosswind of approximately 5 to 10 knots. This is acceptable, but we will keep a close eye on conditions prior to and during landing. We will request braking action reports and closely monitor surface winds during descent and request a wind check prior to touchdown. Decision: Go.

How about an IFR flight in an airplane without ice protection equipment?

Takeoff: We apply the same analysis as for an ice-protected airplane. Decision: Go.

En route: Flying in clouds does not appear to be a viable option. We could fly on top, but we will still have to descend in icing conditions. Any ice would be carried all the way to the ground. Once in flight, even if we attempted to return, we could not reasonably expect to lose all the ice prior to landing. Decision: No-go.

Now let's look at a VFR flight in an airplane without ice protection equipment.

Takeoff: Again, we employ the preceding analysis. Decision: Go.

En route: We will have to stay in VFR conditions. This requires a low-altitude flight below clouds. Ceilings and visibilities are low; however, from the forecast, any precipitation should be in the form of snow.

Personal minimums and training and experience play an important part of this decision. Certainly this would be no place for a student or low-time, inexperienced pilot (although it may be an opportunity for some takeoff and landing practice on a snow covered runway, with a qualified instructor). For a cross-country flight to Wichita, decision: No-go.

An experienced pilot may elect to begin the flight. Here is the dreaded "Let's go take a look." There are no inherent additional risks in this operation as long as we continue to apply sound risk assessment and management to the flight. In addition to those measures mentioned previously, consider flying a major highway, such as an interstate. After a snowfall, remember that the landscape will no longer look like the sectional chart. Many landmarks will most likely be covered with snow. Interstate highways are usually cleared of snow first and, as a last resort, could be used as an emergency landing area. This does, however, eliminate a direct flight to Wichita, but it reduces risk.

Do we have a current sectional chart? These are certainly not conditions to fly with a WAC or outdated sectional. Carefully check for towers or other obstacles en route. Ensure that there are suitable alternate landing fields. Here again, a nondirect route will reduce risk by providing additional suitable alternate landing areas. During the weather briefing, remember to check NOTAMs for all these airports. Decision: Go

En route risk assessment and management become strong players. We are flying from an area of relatively good weather to poor but improving conditions. Should we encounter freezing precipitation, our only option is an immediate retreat. We should always have an alternate in mind, should it become necessary. If the precipitation is snow, do not attempt to penetrate the area unless you can see the other side. Closely monitor the engine for possible carburetor or induction system ice. Check with Flight Watch or an FSS en route for updates, and provide pilot reports.

CASE STUDY

Consider the following: A Cessna 172 pilot received a preflight weather briefing that included marginal VFR conditions and reported icing in clouds near the route of flight. The pilot pulled the carburetor heat control to the on position and descended to about 500 ft AGL to maintain visual contact with the ground. About 1/8 in of ice had formed on the airplane, and the pilot reversed course in an attempt to locate an airport. The flight controls felt "sluggish." The pilot selected a field, configured the airplane, and made a precautionary landing. During the landing, the nose gear sank into the muddy field, then collapsed, and the airplane nosed over.

The NTSB determined the probable cause to be the pilot's continued flight into adverse weather. They sighted as factors the low ceiling, icing conditions, airframe ice, and the muddy field—an alignment of at least four precursors. Unfortunately, this is an example of a situation where the pilot failed to retreat before entering adverse weather conditions. I know it is easy for me to say, but the goal is not to enter these conditions at all.

Landing: Current conditions and forecasts indicate that VFR or special VFR conditions will prevail for Wichita for our arrival time. (Wichita is in Class C airspace; certain Class B airspace prohibits fixed-wing special VFR.) The predominate conditions should be VFR. Here again, this is no place for a student or low-time, inexperienced pilot or a pilot without adequate aeronautical charts. Do we have an alternate, should Wichita fall below special VFR? There are several suitable alternate airports north and west of Wichita. We would not want to proceed beyond these points without positive assurance of landing at Wichita. If we can meet all these requirements, decision: Go.

Thus far we have examined several scenarios for a flight from Goodland to Wichita. What about flying to Wichita from the south, say, Oklahoma City? If this were a morning flight, we would be flying into the teeth of the SIGMET area. If we have ice protection equipment, we might consider a high flight, above the clouds and SLD tops. This would still require a descent into possible severe conditions, for which our aircraft is neither tested nor certified. Therefore, even with ice protection equipment, the best—and safest—decision would be to wait for improving weather, which should only be a few hours away. For alternates, we would again look north, to the area of improving weather. Without ice protection for an IFR or a flight in VFR conditions, decision: No-go. Our best bet would be to wait until the weather system passes.

Flight into Arkansas and the Ohio Valley presents three additional hazards: widespread IFR conditions, high tops, and thunderstorms. With widespread IFR, a suitable alternate may be beyond the range of most small aircraft. Radar shows precipitation tops well above 20,000 ft. Getting on top may not be possible. The AIRMET, neural network, and stovepipe models predict significant icing to 18,000 ft.

Flight to the west, e.g., Denver, would have a different set of hazards. The area is under the influence of upslope, IFR conditions, with snow, and relatively low tops. Aircraft icing is not a serious factor because of the cold temperature. But low ceilings and visibilities and snow-covered runways continue to persist and will continue as long as upslope remains a factor. Whiteout would certainly be a factor in these areas. VFR flight would certainly be out. IFR flight would have to be maintained with conditions close to or below minimums over a relatively widespread area. However, suitable alternates exist to the northeast and east well within the range of most aircraft.

> **ICING MYTH**
>
> I've got full deice; I don't have to worry about any icing conditions.

Absolute instability: A condition of the atmosphere where vertical displacement is spontaneous, whether saturated or unsaturated.

Absolute stability: A condition of the atmosphere that resists vertical displacement, whether a parcel is saturated or unsaturated.

Accretion: the deposit of ice on aircraft surfaces in flight as a result of the tendency of cloud droplets to remain in a liquid state at temperatures below freezing.

Adiabatic process: A thermodynamic change of state with no transfer of heat or mass across the boundaries, where compression always results in heating and expansion results in cooling.

Advection: The horizontal transport of an atmospheric property.

Airmass: A widespread body of air whose homogeneous properties were established while that air was over a particular region of the earth's surface and which undergoes specific modifications while moving away from its source region.

AIRMET Bulletin: An inflight weather advisory program intended to provide advance notice of potentially hazardous weather.

Altimeter setting: A value of atmospheric pressure used to correct a pressure altimeter for nonstandard pressure.

Anti-icing: Prevention of ice from forming on aircraft surfaces.

Atmospheric phenomena: As reported on METAR, atmospheric phenomena consist of weather occurring at the station and any obstructions to vision. Obstructions to vision are only reported when the prevailing visibility is less than 7 miles.

Atmospheric property: A characteristic trait or peculiarity of the atmosphere such as temperature, pressure, moisture, density, and stability.

Augmented: When referring to a surface weather observation, this means that someone is physically at the site monitoring ASOS/AWOS equipment.

AUTO: When used in METAR and SPECI, this indicates that the report comes from an automated weather observation station that is not augmented.

Automated Surface Observing System (ASOS): A computerized system similar to AWOS but developed jointly by the FAA, NWS, and Department of Defense; in addition to standard weather elements, the system encodes climatologic data at the end of the report.

Automated Weather Observing System (AWOS): A computerized system that measures sky conditions, visibility, precipitation, temperature, dew point, wind, and altimeter setting. It has a voice synthesizer to report minute-by-minute observations over radio frequencies, telephone lines, or local displays.

Automatic Terminal Information Service (ATIS): A recorded service provided at tower controlled airports to provide pilots with weather, traffic, and takeoff and landing information.

Berm: A raised bank, usually frozen snow, along a taxiway or runway.

Bleed air: Small extraction of hot air from turbine engine compressor.

Bridging: A formation of ice over the deicing boot that is not cracked by boot inflation.

Calorie: A unit of heat required to raise the temperature of one gram of water one degree Celsius.

Celsius: A temperature scale where 0 is the melting point of ice and 100 is the boiling point of water.

Carburetor ice: Ice formed in the throat of a carburetor due to the effects of lowered temperature by decreased air pressure and fuel vaporization.

Clear ice: A glossy, clear, or translucent ice formed by relatively slow freezing of large supercooled droplets; the large droplets spread out over the airfoil prior to complete freezing, forming a sheet of clear ice.

Closed low: An area of low pressure aloft completely surrounded by a contour.

Coalescence: The merging of two water drops into a single larger drop.

Coefficient of friction: The ratio of the tangential force needed to maintain uniform relative motion between two contacting surfaces (aircraft tires with the pavement surface) and the perpendicular force holding them in contact (distributed aircraft weight to the aircraft tire area). The coefficient is often denoted by the Greek letter (μ). It is a simple means used to quantify the relative slipperiness of pavement surfaces. Friction values range from 0 to 100, where 0 is the lowest friction value and 100 is the maximum frictional value obtainable.

Cold-core low: A low-pressure area that intensifies aloft. When this type of low contains closed contours at the 200-mb level, its movements tend to be slow and erratic.

Conditional instability: A condition of the atmosphere where a parcel will spontaneously rise as a result of its becoming saturated when forced upward.

Conduction: The process of transferring energy by means of physical contact.

Confluence: A region where streamlines converge. The speed of the horizontal flow often will increase where there is confluence. It is the upper-level equivalent of surface convergence.

Contours: Lines that connect areas of equal height on constant-pressure charts.

Constant-pressure surface: A surface where atmospheric pressure is equal—the height of the surface changes with pressure changes, but the pressure itself remains constant.

Convection: The vertical transport of an atmospheric property.

Convergence: An inward flow or squeezing of the air.

Coordinated Universal Time: Formerly Greenwich Mean Time, also known as Z or ZULU time, Coordinated Universal Time (UTC) is the international time standard.

Coriolis force: An apparent force that causes winds to blow across isobars at the surface.

COR: In a METAR/SPECI report this indicates that the originally transmitted material had an error that has now been corrected.

Cumuliform: Describes clouds that are characterized by vertical development in the form of rising mounds, domes, or towers and an unstable air mass.

Cut-off low: See *Closed low.*

Deiced: Ice on a runway has been coated with chemicals.

Density: The weight of air per unit volume.

Dew point: The temperature to which air must be cooled, water vapor remaining constant, to become saturated.

Dew-point front: See *Dry line.*

Diabatic: A process that involves the exchange of heat with an external source, or nonadiabatic; the loss may occur through radiation, resulting in fog or low clouds, or conduction through contact with a cold surface.

Direct User Access Terminal (DUAT):   A computer terminal where pilots can directly access meteorologic and aeronautical information, plus file a flight plan without the assistance of an FSS.

Divergence: A downward flow of air—subsidence—the opposite of convergence.

Dry snow: Snow that has insufficient free water to cause cohesion between individual particles; generally occurs at temperatures well below 0°C.

Dryline: An area within an air mass that has little temperature gradient but significant differences in moisture. The boundary between the dry and moist air produces a lifting mechanism. Although not a true front, it has the potential to produce hazardous weather. It is also known as a *Dew-point front.*

Enhanced infrared (IR) imagery: A process by which infrared imagery is enhanced to provide increased contrast between features to simplify interpretation. This is done by assigning specific shades of gray to specific temperature ranges.

Eutectic temperature/composition: A deicing chemical that melts ice by lowering the freezing point.

Fahrenheit: A temperature scale where 32° is the melting point of ice and 212° is the boiling point of water.

Fall streaks: Fall streaks are ice crystals or snowflakes falling from high clouds into dry air, where they sublimate directly from a solid to a gas.

Flight level: Altitudes flown with the altimeter set to 29.92 in Hg.

Freezing level: As used in aviation forecasts, the level at which ice melts.

Freezing-point depressant: A fluid that combines with super-cooled water droplets forming a mixture with a freezing temperature below the ambient air temperature.

Front: A boundary between air masses of different temperatures, moisture, and wind.

Frontal zone: See *Front*.

Glaciation: The transformation of liquid cloud particles to ice crystals.

Global circulation: See *Planetary scale*.

Geostationary Operational Environmental Satellites (GOES):   An earth-orbiting satellite normally located about 22,000 nautical miles above the equator at 75°W and 135°W. The satellites provide half-hourly visible and infrared imagery.

Gravity wind: See *Drainage wind*.

Ground icing: Structural icing that occurs on an aircraft on the ground, usually produced by snow or frost.

Great Basin: The area between the Rockies and the Sierra Nevada Mountains, consisting of southeastern Oregon, southern Idaho, western Utah, and Nevada.

Heat: The total energy of the motion of molecules, with the ability to do work.

Heat capacity: The amount of heat required to raise the temperature of air, or the amount of heat lost when air is cooled.

Hectopascal: The international unit of atmospheric pressure is the hectopascal (hPa), equivalent to the millibar (mb).

High: Area of high pressure completely surrounded by lower pressure.

Horizontal extent: The horizontal distance of an icing encounter.

Hydroplaning: A condition where a thin layer of water between the wheel and runway causes the tires to lose contact with the runway.

Ice: The solid form of water consisting of a characteristic hexagonal symmetry of water molecules.

Ice protection equipment: Equipment required for an aircraft to be certificated for flight into known icing conditions.

Icephobic liquid: A spray that reduces the adhesion of ice to the deicing boot surface, improving deicing.

Inches of mercury (Hg): For aviation purposes, we commonly relate atmospheric pressure to inches of mercury (in Hg)—altimeter setting.

Infrared: Satellite imagery that measures the relative temperature of clouds or the earth's surface.

Instrument Flight Rules (IFR): FARs that govern flight in instrument meteorologic conditions—flight by reference to aircraft instruments.

International Standard Atmosphere (ISA): A standard reference of temperature and pressure with a lapse rate of approximately 2°C per thousand feet.

Inversion: A lapse rate where temperature increases with altitude.

Isobars: Lines connecting equal values of surface pressure.

Isohumes: Lines of equal relative humidity.

Isopleths: Lines of equal number or quantity.

Isotachs: Used on charts and graphs, lines of equal wind speed.

Isothermal: A constant lapse rate.

Isotherms: Used on charts and graphs, lines of equal temperature.

Jet stream: A segmented band of strong winds that occur in breaks in the tropopause.

Lapse rate: The decrease of an atmospheric variable with height, usually temperature.

Latent heat: The amount of heat exchanged (absorbed or released) during the processes of condensation, fusion, vaporization, or sublimation.

Level of free convection (LFC): The point where a parcel becomes saturated and upward movement becomes spontaneous.

Lifted condensation level (LCL): The level, or altitude, where a lifted parcel becomes saturated.

Liquid water content (LWC): The total mass of water contained in all the liquid cloud droplets within a unit volume of cloud. Units of LWC are usually grams of water per cubic meter of air (g/m^3).

Longwave: See *Rossby waves.*

Low: An area of low pressure completely surrounded by higher pressure.

Low-level wind shear: Wind shear that occurs within 2000 ft of the surface.

Mean effective diameter (MED): The droplet diameter that divides the total water volume present in the droplet distribution in half; half the water volume will be in larger drops and half in smaller drops.

Median volumetric diameter (MVD): The droplet diameter that divides the total water volume present in the droplet distribution in half; half the water volume will be in larger drops and half in smaller drops.

Melting level: The temperature at which ice melts, often referred to as the *freezing level.*

Micron (μm): One-millionth of a meter.

Millibar: Measurement of atmospheric pressure, equivalent to the hectopascal.

Mixed ice: Rime ice and clear ice (mixed ice) are a hard, rough, irregular, whitish conglomerate formed when supercooled water droplets vary in size or are mixed with snow, ice pellets, or small hail.

Mixed icing conditions: An atmospheric environment where supercooled liquid water and ice crystals coexist.

Moisture convergence: An objective analysis field combining wind flow convergence and moisture advection. Under certain circumstances, this field is useful for forecasting areas of thunderstorm development.

Negative vorticity advection: Area of low values of vorticity producing downward vertical motion.

Neutral stability: An atmospheric condition in which after a parcel is displaced, it remains at rest—even when the displacing force ceases.

NEXRAD: The next-generation Doppler weather radar system.

Nucleation: In meteorology, the initiation of either of the phase changes from water vapor to liquid water or from liquid water to ice.

Orographic: A term used to describe the effects caused by terrain, especially mountains, on the weather,.

Overrunning: A condition in which airflow from one air mass is moving over another air mass of greater density. The term usually applies to warmer air flowing over cooler air, as in a warm frontal situation. It implies a lifting mechanism that can trigger convection in unstable air.

Parcel: A small volume of air arbitrarily selected for study; it retains its composition and does not mix with the surrounding air.

Patchy conditions: Areas of bare pavement showing through snow- and/or ice-covered pavements. Patches normally show up first along the centerline in the central portion of the runway in the touchdown areas.

Partial obscuration: In MRTAR/SPECI this is reported when between one-eighth and seven-eighths of the sky is hidden by a surface-based obscuring phenomenon.

Positive vorticity advection: Areas of high positive values of vorticity producing upward vertical motion.

Precipitation: Any or all of the forms of water particles, whether liquid or solid, that fall from the atmosphere and reach the ground.

Pressure: Force per unit area.

Rapid Update Cycle (RUC):   A high-speed computer model that updates every 3 hours, designed for short-term forecasting.

Relative humidity: The ratio, expressed as a percentage, of water vapor present in the air compared with the maximum amount the air could hold at its present temperature.

Ridge: An elongated area of high pressure.

Rime: A white or milky and opaque granular deposit of ice.

Rime ice: A rough, milky, opaque ice formed by the instantaneous freezing of small, supercooled droplets as they strike the aircraft.

Runback: Icing that occurs when local heating of accumulated ice melts, water runs back to unheated areas, and refreezes.

Saturated adiabatic lapse rate: The rate at which saturated air cools or warms when forced upward or downward.

Severe thunderstorms: A thunderstorm that produces winds of 50 knots or greater or hail 3/4 in in diameter or greater.

Shortwave: A trough or ridge embedded in a longwave that moves through the longwave at the same speed as the overall wind flow.

Showers: Showers are characterized by the suddenness with which they start and stop and rapid changes of intensity.

SIGMET: A significant meteorologic advisory that warns of phenomena that affect all aircraft.

Slant range visibility: The visibility between an aircraft in the air and objects on the ground.

Slush: Snow that has a water content exceeding its freely drained condition such that it takes on fluid properties (e.g., flowing and splashing).

SPECI: A special surface aviation weather report.

Squall line: An organized line of thunderstorms.

Snow: A porous, permeable aggregate of ice grains that can be predominately single crystals or close groupings of several crystals.

Stability: The property of an airmass to remain in equilibrium— its ability to resist displacement from its initial position.

Stagnation point: The point on a surface where the local free stream velocity is zero; the point of maximum collection efficiency for a symmetric body at zero degrees angle of attack.

Stratiform: Clouds of extensive horizontal development and a stable air mass.

Stratosphere: The layer of the atmosphere above the tropopause.

Sublimation: The process by which ice changes directly to water vapor or water vapor directly to ice. The sublimation of ice to vapor is a cooling process, and the sublimination of water vapor to ice is a warming process.

Subsidence: Downward vertical motion of the air.

Supercooled: Liquid water or water vapor that exists at temperatures below freezing is called *supercooled.*

TAF: Terminal Aerodrome Forecast.

Temperature: A measurement of the average speed of molecules.

Total air temperature: The result of ambient air temperature and aerodynamic heating.

Towering cumulus: Growing cumulus cloud that resembles a cauliflower, but with a top that has not yet reached the cirrus level.

Tropopause: The boundary between the troposphere and the stratosphere, typically an isothermal layer.

Troposphere: The lower layer of the earth's atmosphere containing about three-quarters of the atmosphere by weight.

Trough: An elongated area of low pressure.

Upslope: The orographic effect of air moving up a slope, which tends to cool adiabatically .

Vertical motion: Upward or downward motion in the atmosphere.

VIP: Video Integrator and Processor used in six levels to indicate precipitation intensity on the radar summary chart.

Virga: Rain that evaporates before reaching the surface.

Visible moisture: Moisture in the form of clouds or precipitation.

Visual flight rules (VFR):   FARs that govern flight in visual meteorological conditions—flight by reference to the natural horizon and surface.

Vort lobe: A contraction for *vorticity lobe.* It usually applies to the 500-mb level and identifies an area of relatively higher values of vorticity. It is synonymous with shortwave trough or upper-level impulse. Generally speaking, there is rising air ahead of the vort lobe and sinking air behind.

Vort max: A contraction for *vorticity maximum.* It usually applies to the 500-mb level and refers to a point along a vorticity lobe where the absolute vorticity reaches a maximum value.

Vorticity: A mathematical term that refers to the tendency of the air to spin; a vertical motion producer.

Warm air advection: A condition in the atmosphere characterized by air flowing from a relatively warmer area to a cooler area. It is often accompanied by upward vertical motion that in the presence of sufficient instability leads to thunderstorm development.

Warm nose: A prominent northward bulge of relatively warm air.

Water vapor: The invisible water molecules suspended in the air.

Wave: A pattern of ridges and troughs in the horizontal flow as depicted on upper-level charts. At the surface, a wave is characterized by a break along a frontal boundary. A center of low pressure is frequently located at the apex of the wave.

Wet ice: An ice surface covered with a thin film of moisture caused by melting, insufficient to cause hydroplaning.

Wet snow: Snow that has grains coated with liquid water that bonds the mass together but has no excess water in the pore spaces.

Whiteout: An atmospheric optical phenomenon in which the pilot appears to be engulfed in a uniformly white glow. Neither shadows, horizon, nor clouds are discernible; sense of depth and orientation are lost.

Wind chill factor: The cooling effect of temperature and wind.

Wind shear: Any rapid, horizontal, or vertical, change in wind direction or speed.

Index

About the Author

A pilot since 1967, Terry T. Lankford holds single-engine, multiengine, and instrument ratings, as well as an FAA Gold Seal Instructor certificate. As an FAA accident prevention counselor, he earned the Flight Safety Award in 1979. A flight service specialist with a degree in physical science/meteorology, he is a member of the Aircraft Owners and Pilots Association and the National Weather Association. Lankford also contributes articles to pilot periodicals and has written several aviation books, including *Understanding Aeronautical Charts* and *Pilot's Guide to Weather Reports, Forecasts, and Flight Planning*, Second Edition, both McGraw-Hill.